CW00747304

AESTHESIA

posthumanities

CARY WOLFE, SERIES EDITOR

51 Aesthesis and Perceptronium: On the Entanglement of Sensation, Cognition, and Matter
Alexander Wilson

50 Anthropocene Poetics: Deep Time, Sacrifice Zones, and Extinction
David Farrier

49 Metaphysical Experiments: Physics and the Invention of the Universe
Bjørn Ekeberg

48 Dialogues on the Human Ape
Laurent Dubreuil and Sue Savage-Rumbaugh

47 Elements of a Philosophy of Technology: On the Evolutionary History of Culture
Ernst Kapp

46 Biology in the Grid: Graphic Design and the Envisioning of Life
Phillip Thurtle

45 Neurotechnology and the End of Finitude
Michael Haworth

44 Life: A Modern Invention
Davide Tarizzo

43 Bioaesthetics: Making Sense of Life in Science and the Arts
Carsten Strathausen

42 Creaturely Love: How Desire Makes Us More and Less Than Human
Dominic Pettman

41 Matters of Care: Speculative Ethics in More Than Human Worlds
Maria Puig de la Bellacasa

40 Of Sheep, Oranges, and Yeast: A Multispecies Impression
Julian Yates

39 Fuel: A Speculative Dictionary
Karen Pinkus

38 What Would Animals Say If We Asked the Right Questions?
Vinciane Despret

37 Manifestly Haraway
Donna J. Haraway

36 Neofinalism
Raymond Ruyer

35 Inanimation: Theories of Inorganic Life
David Wills

34 All Thoughts Are Equal: Laruelle and Nonhuman Philosophy
John Ó Maoilearca

33 Necromedia
Marcel O'Gorman

32 The Intellective Space: Thinking beyond Cognition
Laurent Dubreuil

31 Laruelle: Against the Digital
Alexander R. Galloway

(continued on page 244)

AESTHESIS AND PERCEPTRONIUM

On the Entanglement of Sensation,
Cognition, and Matter

ALEXANDER WILSON

posthumanities **51**

University of Minnesota Press
Minneapolis
London

Portions of chapter 4 were originally published as "Beyond the Neomaterialist Divide: Negotiating between Eliminative and Vital Materialism with Integrated Information Theory," *Theory, Culture, and Society* 35, no. 7–8 [special issue on the New Materialisms] (2018): 97–116.

Copyright 2019 by the Regents of the University of Minnesota

All rights reserved. No part of this publication may be reproduced, stored in a retrieval system, or transmitted, in any form or by any means, electronic, mechanical, photocopying, recording, or otherwise, without the prior written permission of the publisher.

Published by the University of Minnesota Press
111 Third Avenue South, Suite 290
Minneapolis, MN 55401-2520
upress.umn.edu

ISBN 978-1-5179-0659-7 (hc)
ISBN 978-1-5179-0660-3 (pb)

A Cataloging-in-Publication record for this book is available from the Library of Congress.

The University of Minnesota is an equal-opportunity educator and employer.

UMP LSI

THIS BOOK IS DEDICATED TO THE MEMORY OF MY FATHER, BARRY WILSON

CONTENTS

Acknowledgments   ix

Introduction: Nonhuman Aesthetics   1

1 Chaos and Cognition   19

2 Determinism and Drift   65

3 Aesthesis and Prosthesis   109

4 Superposition and Time   155

Conclusion: Aesthesia in the Wild   207

Notes   217

Index   231

ACKNOWLEDGMENTS

I COULD NOT EVEN HAVE BEGUN to pursue this course of research without the constant support and encouragement of several key people. First, if it weren't for the personal and philosophical generosity of Bernard Stiegler and Brian Massumi, I would never have found the courage. I must extend my sincerest gratitude to all my colleagues at Aarhus University; the comfortable postdoctoral position they offered can be considered among the important "material causes" of this book. In particular, I thank Jacob Wamberg for the many insightful discussions and friendly philosophical debates we had in Aarhus, Johanna Seibt for her constructive critiques of my work as it was developing, and all my AU students who had the courage to delve into these speculative topics with me. I would also like to express my appreciation for the brief yet decisive conversations I had with the following fine thinkers, who each helped me clarify certain points in the book during the writing process: Karen Barad, Michel Bitbol, Gabriel Catren, Quentin Meillassoux, and Arkady Plotnitsky. I thank Cary Wolfe, Douglas Armato, and Gabriel Levin, the editorial board of the University of Minnesota Press, as well as my reviewers, for successfully steering me through. I thank Sara Baranzoni, Joanna Dietzel, Benoit Dillet, Yuk Hui, Iwona Janicka, Anaïs Nony, Erin Sexton, Alexandre St-Onge, Patrick Valiquet, Paolo Vignola, Paul Willemarck, and Maria Witek for the helpful dialogues that, alas, happen all too rarely. Finally, I am deeply grateful to Ricarda Bross for having been so supportive in the final push. Thanks.

*AESTHESIS: experience
in the world*

INTRODUCTION
Nonhuman Aesthetics

WHAT IS REQUIRED OF A CLUMP OF MATTER for it to have an experience of something, for it to have a perspective on the world? The new millennium's theoretical challenges to anthropocentric thought urge us to respond to this question. The realization that our life-sustaining ecosystem has been irreparably transformed by our activities, coupled with the realization of the remarkable fragility of human existence, has challenged even our most common notions of the distinction between nature and culture. The Anthropocene's most difficult theoretical impasse may therefore be its problem of "inclusion": who (or what) should be included in political, ethical, and aesthetic considerations today? Politics has always in some way turned on the definition of the *polis*: its boundaries and its principles of inclusion and exclusion. But today we are pressed to radically widen the scope of politics to potentially include animals, insects, bacteria, weather patterns, computer algorithms, and various other nonhuman entities. Given that our anthropocentric biases are the cause of our impending finality, it is both more urgent and more difficult than ever to decide where to draw that line in the sand.

Is the difference between animal sentience and human sapience a qualitative distinction, or is it rather a question of degree? Can a rock or a table have an experience? In their responses to these questions, the various speculative ontologies being championed in recent years almost invariably fall into one of two camps, two radically incompatible perspectives concerning the place of experience—or *aesthesis*—in the world. There are those who attribute experience to all things (vitalists, panpsychists), and those who are skeptical of the concept of experience, who claim that it lacks cogency and should be abandoned in favor of human-defining notions like knowledge, reason, and intelligence (eliminativists).

· 1

In other words, the dilemma between panpsychism and eliminativism presents two opposing takes on the relation between matter and mind that appear to be symptomatic of the Anthropocene as a period of theoretical crisis: in their extreme forms, one says that aesthesia is everywhere while the other says that aesthesia is nowhere. On the first side, we indiscriminately attribute subjective integration to all objects and materials; on the second, we deny that subjectivity even exists.

Material semiotics, actor-network theory, cyborg feminism, critical posthumanism, vital materialism, and object-oriented ontology are examples of a trend toward radically inclusive frameworks that adopt "flat" ontologies, and therefore fall on the panpsychist side of the dilemma. I am sensitive to such radical inclusions of matters of concern, and I agree that the Anthropocene should be a moment to question our blind faith in the exceptionalist modern (and premodern) understanding of the human being's special status in the universe. However, I am skeptical of the idea that we should be "indiscriminately" pluralistic. It seems theoretically lazy to vaguely attribute some sort of experiential withdrawal to all objects indiscriminately just because we fear undermining or "overmining" their possible interiority or autonomy. But more importantly, such a position of radical inclusion often corresponds to an *apolitical* stance. Considering the feelings of household furniture and of clouds in political discussion severely diminishes the emancipatory quests of those who are actually and expressedly involved in political struggles. In other words, from a political perspective, saying "all objects matter" is just a radical version of the much-criticized "all lives matter" response to the recent Black Lives Matter movement. And in addition to this, since the *polis* necessarily implies a *boundary,* indiscriminate pluralism ultimately fails to resolve the question "what is to be done?" For such stances of absolute indiscriminate inclusion face an infinite regress of considerations: if we must consider *everything* before we take action, when will we ever be in a position to act? We might go on including more potential agents and experiencers forever, in effect delaying action indefinitely. This echoes the "frame problem" in artificial intelligence, a similar problem of consideration of conditionals that leads a program into an endless regress. In our almost paranoid, if commendable, concern for the potential feelings of every parcel of the universe, we seem to be condemned to an endless postponement of the act. Action seems to paradoxically imply that we eventually stop counting, that we inhibit this perpetual iteration of matters of concern, that we look to the horizon and see the forest through the trees. The question is how to define new criteria for deciding, at least provisionally, how, when, and where to draw that illusive line, while nevertheless remaining committed to the anti-anthropocentric stance. The distinction hinges not merely on the capacity for agency, as many ventures into this issue have urged, but on the question of which materials can experience, perceive, sense, or cognize. Somewhat ironically, the possibility of including alien matters of concern into our own sphere seems to actually demand that we learn to draw this distinction, even if only provisionally.

Of course, eliminativists face obstacles of their own. We cannot merely "explain away" first-person experience, *qualia*, affects, intentionality, and so on, just by calling them cognitive illusions. Heat is caused by the micro-movements of energized particles that are not themselves hot. Time is caused by our thermodynamic implementation within the cosmos, which may not itself be temporally asymmetrical. Yet it would be misguided to say that these are illusions; heat and time are *real* phenomena, just as real as the auxilliary phenomena we mobilize to explain them. The realization that the ultimate cause or origin is unthinkable, unknowable, and inadmissible within science means that the playing field is level. Indeed, as part of nature, these appearances should be taken as elements that must be included in our account of, as Wilfrid Sellars put it, how things *hang together* in the broadest possible sense. Both the scientific image and the manifest image must be accounted for within the picture. So, at its worst, when eliminativism simply dismisses the manifest image in favor of the scientific one, it obviously fails to found philosophy in the actual state of affairs.

In their defense, however, some card-carrying eliminativists do not actually just "explain away" experience. And they are correct in their shared general position that people are often too attached to their preconceived "folk" notions of experience and psychological states. Science is all about challenging those notions, and philosophy shouldn't be fighting the science. Rather, it should strive to put it all back together into a broader picture. But many philosophers of mind who defend various flavors of dualism, vitalism, and panpsychism will never be satisfied with any functionalist explanation of the goings-on of experience. They can always retreat further into the backstage recesses of what Daniel Dennett calls the "Cartesian theatre" and say, "ah, but that is just an explanation of the experience, not the experience itself." This is essentially the character of Thomas Nagel's "What is it like to be a bat?"[1] No matter how well you describe the bat extrinsically, you still don't know what it's like to be a bat. But this argument is fundamentally flawed: it confuses what it is to *know* something with what it is to actually *be* something. I side with Paul Churchland here. The reason the explanation of the bat's internal experience does not instantly make you feel or experience what it is like to be a bat is simply that the description is not true of you. "What is required to make you a batstyle cognizer—to make you enjoy the special dimensions of a bat's subjective cognitive activity here at issue—is that the complete physical theory of batstyle cognition *be true of you*."[2] This kind of reasoning does not mean that experience does not exist. It indeed exists just as much as heat and time do, but it requires situating within the big picture. How can a dimension of private experience emerge from a material substrate that would seem to not have such a dimension at all?

Taken together, the following chapters can be said to sketch out a pragmatic way forward, a middle way between panpsychism and eliminativism. As we will see, my own reflections lead to an understanding that is in line with the vitalist and panpsychist's assertion that aesthesia is not restricted to humans or even to higher organisms, but potentially extends much further into the material world. However, simultaneously, I side with the rationalist, functionalist, and eliminativist in arguing that our traditional notions of experience, sensation, *qualia*, givenness, and subjectivity are too often unconvincingly vague and unfounded. Insisting that experience must be inscribed in the deterministic toils of matter, I reject the dualist overtones of vitalism, and argue that a theoretical response to the dilemma should be able to describe, as rigorously as possible, the criteria allowing us to discriminate between those materials and objects where aesthesis takes place and those where it does not. Simply blanketing all of reality with a vague sense of "maybe it is conscious," I believe, does no one any favors.

Developing a theoretical middle way will require accounting for and assessing the relation between the traditional categories of aesthesis and noesis after the historical collapse of foundationalism. The epistemological relevance of subjective experience and of foundationalist accounts of knowledge has been radically put into question as much by developments intersecting the analytic and pragmatic strands of philosophy as it has been by those in the continental tradition, leaving aesthetics as the science of the "lower cognitive faculty" of sensation—as it was originally defined by Alexander Gottlieb Baumgarten—difficult to legitimize. Indeed, developments in science, mathematics, and epistemology, as well as in the arts, have echoed the same realization: the long-held prejudice of a strict demarcation between sensation and knowledge, between sentience and sapience, has been discovered to have been unfounded. Reimagining aesthetics beyond the human today converges with a coming to terms with the inherently entangled nature of sensation, cognition, and matter. So, by recasting aesthetic questions within the framework of what I call "epistemaesthetics," which considers knowledge and aesthesia as a single category of concern, I propose a speculative theory of nonhuman experience that avoids the pitfalls of both indiscriminate pluralism and eliminativism, while remaining steadfastly committed to materialism, naturalism, and the "scientific image."

As living beings, we are irreducibly perspectival, asymmetrical, and oriented within the broader structures and processes of reality. In a tradition inaugurated with Kant's critiques, philosophy has been concerned with describing the conditions of this subjective experience and the possibility of knowledge. There is a background to all thought, for each event of cognition is necessarily enabled by processes that it simultaneously eclipses. What we experience, how we cognize, is always conditioned by a deep structure of nested transcendental constraints, and these constraints evolve in time: they have a history. Furthermore, these constraints may not be the same in all contexts: the event of cognition is no doubt somewhat different in each case, from concept to concept, individual to individual, culture to culture, species to species, and kingdom to kingdom. But it is never enough to merely intuit these constraints in a priori thought: in order to give an account of their genesis and of their diachronic and synchronic variety, we must strive to give a naturalistic account of what it is in the toils of the material world that produces these constraints on cognition, human or otherwise. Here my approach might be said to echo Gilles Deleuze's *transcendental empiricism,* the goal of which is to describe the conditions of *real experience,* rather than merely *possible* experience. Following Solomon Maimon, it became clear quite soon after the introduction of Kant's critical project that, in a sense, Kant was not being critical enough. What was missing was a critique of the implicit assumption that the conditions, faculties, and categories of experience are fixed and implacable for all cognizers, in all instances. This underexamined assumption of the fixity of human finitude was inseparable from the notion that the conditions of possibility of thought could be distilled from pure a priori reasoning. But the very notion that necessary truths can faithfully be gleaned a priori itself depends on the assumption that the given insights would be the same in each case of reflection, that they could not have been otherwise, according to fixed transcendental categories. Thus we smuggle in the purity and necessity of our categories through the assumptions of the operation of a priori synthesis. But there is no such pure and fixed form of a priori reasoning: the a priori thought is a thought that effaces its own conditions and assumptions. It masks its origins and traffics in tautological presupposition, leading us to believe that it emerges spontaneously from the pure constraints of the real, when in fact it too must have a history, a contingent path receding into indistinction. Whatever categories we may intuit as *necessary conditions* of experience will always themselves be contaminated by the *contingent conditions* of our reasoning.

Given these unavoidable constraints on subjectivity, one way to understand the *telos* of philosophy is as a process of *progressively* pushing back the boundaries of our cognitive blind spots and horizons, reconciling and integrating more and more irreducible perspectives upon the transcendental complex that constitutes us, thereby increasing our "degrees of freedom" within it. The nested biological structures and evolutionary constraints that cooperate in the production of each waking human experience themselves arise from the contingent integration of prior independent perspectival processes, each new emergent whole sublating and supervening on the previous ones. As we study any aspect of science, whether it be the evolutionarily forwarded psychological biases that skew our evaluations and intentions or the strange behaviors of quantum physics that defy our common-sense of objectivity, we are indeed shedding light on some of the transcendental conditions of our subjective experience. By progressively naturalizing transcendental constraints on subjectivity, explaining biases and perspectives through coherent logical, empirical, and historical accounts of their genesis, we thereby expand our degrees of freedom within these very constraints.

Of course, as a matter of principle, cognition will never fully *transcend* its perspectival predicament, since it is bounded by definition. An organism or intelligence cannot surpass its own boundedness, at least not synchronically. In a diachronic sense, we can assume that each paradigmatic shift in our models of objectivity, each transcendental type-defining transition, is a "transcending" of the prior model's conditioning constraints on our capacities to experience, but the synchronic horizon is constitutive of subjectivity, of knowledge, of reason, and of the choice between this or that, the decision of a world. To frame this in the terms of the new realisms, in my view, the choice should not be between "correlationist" and "anti-correlationist" accounts, but rather between accounts that take a given form of correlation or finitude as complete, static, or unproblematic and those that take our current correlations as materials that may be challenged, transformed, hybridized, expanded, or collapsed. Here I agree with Gabriel Catren: "The fact that human experience is necessarily framed by a system of transcendental structures (physiological, technological, conceptual, linguistic, cultural, and so on and so forth) does not imply that we cannot try to modify, deform, or perturb these structures."[3] Realizing that subjectivity is necessarily conditioned by forces beyond its current purview is no reason to renounce altogether the project of expanding our degrees of freedom within these structures, processes, and influences. For indeed, if we fail to explore and integrate variations and limit cases of cognition, we fall prey to the same transcendental error that was Kant's: to assume that our place within these conditioning factors is fixed and permanent. Thus, it is not a matter of taking sides between correlationism and anticorrelationism: realism implies a rigor of thought that accepts both its finitude and its incompletion and embraces the indistinction that swarms around the intelligible. Even though we cannot step out of subjectivity, we must nevertheless not resign ourselves to remaining in a fixed form of human subjectivity.

Having this in mind allows us to shed light on the question of *nonhuman aesthetics,* a project to which I wish to contribute with this book. For, from this perspective, it seems clear that the effects of the Anthropocene period greatly challenge anthropocentric thought and that this challenge converges with the prospect of posthumanist enlightenment: the critique of human fixity and its associated hybridization with nonhuman forms of subjectivity turns out to be congruent with the advancement of knowledge and the destiny of ideas. Deleuze sometimes defined philosophy as the project of *avoiding stupidity,* and I believe this definition applies quite well in this context. To avoid stupidity, to avoid remaining "all too human," we must pursue the expansion of our phase spaces by perturbing our frames and sublating seemingly incompatible perspectives into our "total science." This goes hand in hand with uncovering the processes at work in the co-constitution of subjectivities and their correlated objectivities, or *Umwelt*s. It is a matter of operating deliberate variations on cognition, exploring and sometimes exceeding its current limits, and integrating the seemingly incompatible perspectives and worlds these variations reveal into an ever broader, more inclusive framework.

One way to approach the reading of this book, therefore, is as a sketch of such a provisional naturalizing account of the genesis of the constraints on forms of experience and cognition. The book provides a flexible, speculative report of how various nested processes and causal constraints in multiple domains of consideration might participate in the constitution of subjectivity (and its correlated objectivity). Each chapter of the book progressively widens the scope of the inquiry. Starting with a focus on human cognition, we move on to questions from biological and technological evolution, then on to discussions of how aesthesis relates to the emergence of complexity in nature, and finally on to ontological and cosmological considerations. Eventually, the investigation will carve out an account of *aesthesia as a state of matter* and I will offer speculations as to the conditions of "perceptronium." I borrow this convenient term from a certain underappreciated quantum physical construal of consciousness,[4] slightly modifying it to refer to the simplest material system corresponding to some nonzero amount of aesthesia. Does perceptronium require a brain? A body? A nervous system? Does it need to be *alive* in the traditional sense of the word? There is perhaps an infinite regress of contextual frames that constrain the possibilities of experience, and of course this book can only scratch the surface of the topic. Nevertheless, I offer speculative responses to these questions by discussing how aesthesis relates to current accounts of the emergence and evolution of life, to hominization, to technogenesis and artistic practices, and to cosmogenesis more broadly.

Chapter 1, "Chaos and Cognition," establishes the conceptual lens for the rest of the book. In light of findings in multiple domains of thought in the last two centuries pointing to the inherent entanglement of knowledge and sensation, I will argue that the categories of epistemics and aesthetics should be replaced with a single framework: epistemaesthetics. Science and philosophy in the West have historically considered knowledge as an opposition to the senses: knowledge has been held to go above and beyond mere feeling or sensation. According to this account, though we may say in a vulgar sense that an animal "knows" its environment, it certainly does not "know that it knows." In the last century, parallel developments in several different disciplines and traditions have suggested that it is impossible to formally disentangle the higher and lower "cognitive powers" on which the distinction between sensation and knowledge traditionally rested. This insight follows from the realization that synthetic and analytic forms of knowledge cannot be formally distinguished because they are hopelessly intertwined, as are accidental and existential

philosophy

properties and necessary and contingent truths. From Bertrand Russell through Ludwig Wittgenstein, Rudolf Carnap, Wilfrid Sellars, and Willard Van Orman Quine, and from Edmund Husserl through Martin Heidegger, Jacques Derrida, and Deleuze, philosophy on both sides of the divide has been urged toward a consensus that foundationalism, the idea that justifications for knowledge can be neatly reduced into primary sensations or nonpropositional engagements with reality, has failed. The history *math* of mathematics has followed a similar path: from the discovery of non-Euclidean geometries and the paradoxes of Georg Cantor's set theory to Kurt Gödel's discovery of the necessary incompleteness of formal mathematical systems and Alan Turing's confirmation that many mathematical proofs are incomputable. And the theories of science have also taken *science* similar steps in the last century. While Karl Popper attempted to legitimize science by defining it in terms of deduction and falsification rather than induction and verification, the Duhem-Quine thesis, which exposed the arbitrary nature of how scientists choose which assumptions to modify in the face of contrary evidence, ultimately dissolved Popper's strict *demarcation*. The field has realized that the apparent necessary truths the sciences uncover actually derive from contingent axiomatizations that, though not arbitrary, derive in turn from human cognition's specific path of development within its environment.

These realizations obviously pose important challenges to aesthetics as the science of a separate "lower power" of cognition. It is all the *Aesthetics* more surprising that aesthetics still echoes this prejudice, given that artists themselves were among the first to figure out that they could not fully disentangle the two levels: perceiving and knowing. Indeed, if Western art's conquest of illusionism lost its raison d'être after the impressionists, it was not due to the invention of photography or some other outside influence; rather it was that artists themselves came to realize the impossibility of what John Ruskin had called the "innocence of the eye," meaning the isolation of seeing from knowing, the purity of appearances rid of all infections by epistemic biases and conventions. There can be no innocent eye. Seeing, perceiving, feeling—sensing in all its guises—is always already imbued with knowledge, with conceptual commitments, rationalizations, intentionality, and so on. Reinventing aesthetics today requires finally coming to terms with the radical implications of this realization: I believe a shift in focus is required that responds to the apparent dissolution of the strict demarcation between the lower and higher cognitive powers, sentience and sapience. The infant, the brute, and the living cell must be taken to already be constituted by epistemaesthetic entanglings: as Jacob Von Uexkull had already discovered, there is no point at which life simply perceives without bias, and this bias is none other than the organism's functional instantiation, which is essentially a form of material memory of what happens and operational knowledge of how to update its horizon of expectation given certain perturbations.

So what might the term "aesthetics" imply now that there seems to be no way out of the "web of beliefs," no way "outside the text," now that it seems that knowledge never finds its way down to pure sensations and that experience is always through and through suffused with epistemic biases and inclinations? Ironically, it is an oft-overlooked aspect of Leibniz's initial insights with respect to aesthetics, on which Baumgarten founded the discipline, that will allow us beyond this impasse. Misunderstood and subsequently all but abandoned by the discipline, his subtle linking of the indistinction of *qualia* with mathematical uncountability actually already suggested the need to understand cognition and aesthesis as a single category. We will hence revisit some key moments in the history of the notion of indistinct knowledge in relation to randomness, contingency, and the constitution of intelligibility. Through a discussion of several concrete examples from science, philosophy, art, divination, and the theory of computation, chapter 1 offers a way of thinking about the relation between knowledge and perception, between noesis and aesthesis, that does not draw a hard line between them, but rather suggests that their associated phenomena concern respectively different aspects of the complex world we live in, and about how the constraints on subjectivity offer the world to experience. If what we have traditionally referred to as *aestheta* are indistinct, ineffable *je ne sais quoi*s of experience, they may not be completely divorced from the web of beliefs that constitutes distinct knowledge and the conditions of intelligibility, for it too has been shown to, around the edges, recede into confusion. In other words, *aestheta* can be interpreted as real aspects of the (transcendental) structures that constrain the evolving relations between subjectivities and their worlds, locations where these configurations act upon each other, neutralizing or interfering with each other, rather than where they terminate.

Again, such transcendental influences on subjectivity must be interpreted naturalistically as material, biological, cultural, and technological in character: they are real features of our world that act upon the constitution of subjectivities; they are constraints that we necessarily rub up against whether we test a scientific hypothesis or take in the beauty of a foreign landscape. All knowledge, all meaning, all experience, derives from these path-dependent interactions, these entanglements between organism and world. So, between *noeta* and *aestheta* is not a hard distinction, but rather a spectrum of varying textures, reflecting the fluctuating features, interactions, and complex interferences between the sedimentations and assemblages that the long path of our organismic development has secreted in its wake, and that still hold sway on our capacities to know and perceive. Importantly, this reassessment of *aestheta* with regard to *noeta*, I believe, speaks directly to the Anthropocene-period plea for considering forms of nonhuman or alien aesthesis, allowing more rigorous explorations of the degrees of freedom of subjectivity, both possible and compossible: organisms following different phylogenetic paths will have evolved different constraints; their forms of subjectivity will be incommensurable with each other.

In the realm of human practices, there is an echo of the classical philosophical prejudice in favor of distinct knowledge: aesthetic pursuits are often deemed socially trivial and generally irrelevant in comparison to technological or survival-related activities. This prejudice too is founded in the traditional distinctions between necessary and contingent, essence and existence, or analytic and synthetic. Just as science and philosophy have historically privileged the role of distinct knowledge over confused knowledge, I contend that an analogous prejudice upholds the virtues of technology with regard to the arts and aesthetic practices. Hence, I argue in chapter 2, "Determinism and Drift," that our tendency to interpret history as being driven by technical determinism and to champion what is explicitly deemed "useful" is correspondingly misguided. The useful is always determined through contingent operations, hinging on auxiliary assumptions that are often not themselves made explicit. And there are salient examples implying the contrary. As is suggested by the archeological site of Göbekli Tepe, for instance, the transition from nomadic to sedentary life may have been motivated by aesthetic pursuits rather than technical ones. The invention of photography also seems to have been motivated not by technological affordances proper, which were already present hundreds of years prior, but by factors that escape the purely technological domain. But this is not to say, of course, that technical and material determinants do not play a crucial role. They do. Indeed, all constraints are material constraints: even the most spiritual or logical operations must happen in and through matter, if they *happen* at all. It is rather that, as has been observed by the recent "new materialisms," matter is not

New
materialisms

as stable and inert as we formerly thought. Though I cannot support such recent theses that consider matter in vitalist terms, it is nevertheless true that matter has shown itself in the last century to be a principle of unrest, contingency, randomness, and fluctuation. And technology echoes this with its essential *pharmacological* nature: it is undecidable and cannot be tamed by our provisional distinctions between useful and useless. Panglossian reductions to *purposes* simply redouble the fallacy of thinking matter as a set of immutable attributes.

This is reflected in the widespread prejudice for privileging technical affordance over aesthetic concerns at the heart of the only "scientifically sanctioned" discipline of nonhuman aesthetics: evolutionary neuro-aesthetics. In addition to anthropocentrically limiting the scope of nonhuman aesthetics to the human's evolutionary ancestry, the neuro-aesthetic paradigm privileges technical affordance by connecting aesthetic preferences to sexual selection. As we will see, this distinction between environmental selection and sexual selection simply mirrors the distinction between the technical and the aesthetic. The tendency to downplay the significance of sexual selection may have originated in its "Wallacean" interpretation. In the original Darwin–Wallace debate, Alfred Russell Wallace argued that sexual selection was only a trivial part of the evolutionary story because, ultimately, sexual preferences too were determined by environmental factors. From neo-Darwinism on to signaling theory, environmental selection has since been considered more significant than sexual selection, just as the technical is deemed more "useful" and determining than the aesthetic. I argue, however, that many such evolutionary narratives about the place and purpose of aesthetics are in fact what Stephen Jay Gould called "just-so stories," in reference to Kipling's tales about how various beasts acquired their traits. In other words, they are unfalsifiable and unverifiable fictions that merely reaffirm cultural biases that privilege one term over the other. With support from several well-founded critiques of adaptationism, the chapter thus deconstructs the method of "reverse engineering" that guides much of the thinking in this field.

Furthermore, such views now face important challenges by competing interpretations of the place of aesthetics in evolution. Flipping Wallace's argument on its head, we will consider an alternative theory that revives the work of Ronald Fisher and advocates that sexual selection should be understood as a variant of "aesthetic selection." It will be suggested that aesthetic selection happens in equilibrium situations, zones of underdetermination by evolutionary forces—that is, where the combined effects of material constraints are reciprocally neutralized, resulting in what statisticians call a "null model," where one cannot gauge whether a phenomenon is determined by a specific individual cause or random (or determined by a nexus of infinite causes). There is a compelling parallel between this construal of aesthetic selection and the original Leibnizian idea that aesthetics concerns the indistinct, which results from the reciprocal neutralization of multiple forces. I am inclined to think that aesthetics should be rehabilitated in somewhat the same way with regard to technology. I therefore apply this insight to the related notion that art originated in the territorial animal, a theme running through the philosophy of Étienne Souriau, Deleuze and Guattari, and Elizabeth Grosz. I challenge this idea by showing that territoriality can be thought of as equivalent to the Heideggerian conception of technology as "enframing." By contrast, I argue that we should think of art as concerning regions outside or in between the technical "frames": just as aesthetics concerns the indistinct, the underdetermined, and the paradox that conditions intelligibility, rather than being allied to the territory as such, the artistic act concerns the gap or transition between territories.

After having resituated aesthetic pursuits with regard to the technical in the cultural and evolutionary sphere, it is possible to further generalize aesthesis for the wider nonhuman material context. As we will explore in chapter 3, "Aesthesis and Prosthesis," the emergence of complexity in nature presupposes processes identical to those that allow us to distinguish between art and technè in the human domain. Gilbert Simondon characterized aesthetics as a field of "reticulation" that unifies and produces reciprocal relations between otherwise disparate elements of the human life-world. But, when we confront Simondon's description of general patterns of individuation, from crystals to biology to technological formations and psychosocial entities, with Bernard Stiegler's conception of mnemonic exteriorization, it becomes possible to greatly generalize aesthesis as a process for the nonhuman domain. What I call "general exteriorization" involves two complementary factors: intercession (aesthesis) and supersession (prosthesis). I argue that supplementarity and prosthesis, notions characteristic of the philosophy of mnemotechnics in Derrida and Stiegler, are actually also typical of all living systems, or rather, all *extended critical systems* (a concept I borrow from Guiseppe Longo and Francis Bailly).

Stiegler observes that technologies form a "tertiary level" of retention, further exteriorizing the Husserlian levels of primary and secondary

concrescence - the process of the many becoming one

mnemonic retention. But I show that, even within human physiology, we observe many more nested levels of retention and protention. For instance, the human brain's interface with the world is already *mediated* by layers of neurological "exteriorization"; each cortical layer is a successive prosthetic exteriorization of the previous, constituting what is effectively the organic analog of technical supplementation. The logic of the supplement was at work when the unicellular organism evolved the "technology" of the flagellum: the flagellum *supplements* the organism's capacities, both augmenting and replacing them, just as the adoption of new technologies supplements human capacities. Organisms are always already in some sense "proto-technical." But like prosthesis, aesthesis too can be observed in the wider nonhuman context: taken in the Simondonian sense of *aesthetic reticulation,* aesthesis can be said to correspond to the process of integration on which emergence hinges. The parts of a system need to integrate synergistically for new effects to appear on the level of the whole. The observation that aesthesis complements the process of organic prosthesis at every step suggests how it may be resituated in the broader material context. Each emergent level of complexity depends on the reticulation of the parts (intercession), which are synergistically bootstrapped into properties on the level of the exteriorized whole (supersession). I will suggest that aesthesis can be construed, in the most abstract sense, as a material process of synthesis or integration that also corresponds to what Alfred North Whitehead called "concrescence," the process of the "many becoming one." The successive levels of emergent complexity, each supervening upon those below and exhibiting irreducible behavior, I would like to suggest, correspond to the nested paired events of intercession and supersession, synthesis and analysis, reticulation and discretization, or again, aesthesis and prosthesis that characterize emergence in nature.

Only after having expanded the scope of aesthesis in this way can we turn to ontological and cosmological considerations, as we do in chapter 4, "Superposition and Time." The question of the relation between cognition, emergence, and nonlinear phenomena obviously spills over into considerations about the nature of time and causality. Such considerations are moreover motivated by the recent revival of realism in contemporary discourse. One familiar case is that of Quentin Meillassoux's critique of Kantian "correlationism," mentioned above, and his accusation that all such thinking is ultimately solipsistic. I juxtapose Meillassoux's account with Michel Bitbol's defense of correlationism to show how these seemingly contradictory positions actually exhibit a strange symmetry that renders them almost assimilable. Together, they reveal very interesting perspectives on the place of experience or perception in the material world. As we will see, their diffractive encounter suggests that it is possible to consider aesthesis not merely as a process existing in time, but as a *condition* for time and causality. This claim is warranted by virtue of the fact that the irreversibility of time depends on the "coarse-graining" of nature. As Pierre-Simon Laplace observed, if one had knowledge of all the causes of all the effects in the universe, the "future as the past would be present" to one's eyes. The irreversibility of time is therefore inseparable from the necessarily incomplete character of knowledge. The fact that we do not know the causes and cannot trace the individual microscopic details of the universe explains this irreversible character of time. And yet, there can be no world that is not expressed in specific perspectives: to be a world, the possible must be rendered asymmetrical, skewed, selected. The second law of thermodynamics, "time's arrow," which forever proceeds toward increasing entropy, is furthermore inseparable from this aesthesic coarse-graining of the universe. Indeed, entropy corresponds to *hidden information*: information about the microstates of a system that we cannot recuperate or track individually, such as the individual trajectories of the particles making up a cloud. Time's arrow is an effect of nature's coarse-grained revealing: if we had access to all the individual events leading through causal chains to and from the given moment, there would be no time. But since, as I will argue, the coarse-graining of nature is an integral aspect of the process of aesthesis, the ontological reification of the distinction on which reductionism hinges will seem, again, quite misguided. Aesthesis does not choose between subjectivity and objectivity, for it produces both aspects simultaneously: the ungrounding of matter and the constitution of perspectives on the world are two facets of the same operation.

Again, the rigor of thought requires the acceptance of its incompletion. And this, as we will see, is perhaps the great irony of aesthesis: the world would not coarse-grain itself before us were we not to take it seriously. If we disbelieved in the real, if we isolated ourselves in a solipsistic bubble that extended only as far as what could be grasped within the finitude of our mind's eye, never would we encounter the indistinct; never would we have to accept the contingency of it all, the chaos and spontaneity behind it all. Our pseudoworld would be populated exclusively with finite concepts, with distinct types and their tokens. But a

In the right margin, handwritten: *Diffractive encounter*

commitment to realism, to the belief that the world contains real problems with which we are engaged and that these are not just figments of the imagination, means accepting the reality of what the type theorist calls the "empty type." As demonstrated by the common history of typed systems in mathematics and functional programming languages in computer science, we can in specific circumstances tidily do without the consequences of the incomputable and conveniently restrict ourselves to those concepts that we can construct and capture within the finitude of thought. But it would be tantamount to solipsism to transform this practice into an ontological commitment, to reject the autonomy of the indistinct and the irrational like Kant or Descartes. The empty type is not an empty set: though it is devoid of determinate elements, it is filled with the very procedures through which escapes from the correlation are continually performed.

Realism implies the *chaosmos*: an inclusion of distinct and indistinct. Ouroboros chases its tail as long as we deny the indistinct. This is reflected in all answers to the question "why is there something rather than nothing?" that has motivated cosmological arguments for the *first cause* and the *unmoved mover* from time immemorial. Modern cosmological physics encounters a similar question: why was the universe in such a low entropy state to begin with? In other words, why was Humpty Dumpty sitting on the wall in the first place? The universe could indeed have been inaugurated with him already lying broken on the ground, and there would have been no fall, no need for causation of any kind. With the help of Deleuze's concept of *quasi-causality*, I show that the process of aesthesis in fact presupposes this lower-entropy past. Applying *quasi-causality* to the history of work on the so-called "anthropic principle" in cosmology, and the related notion of "observation bias," I show that this leads to an apparent loop between perception and the state of the universe where, as physicist John Wheeler noted, cognition *participates* in the unfolding of the cosmos. If observers are material things, matter is made of information, and information is founded in the activities of specific observers, then we find that distinct cognition and the universe are bound in a logical loop, like Ouroboros chasing its tail. But, if realism implies breaking this circle open, it also implies a modification of the tidy interpretation of matter as information. Echoing Meillassoux's account of "speculative materialism," matter cannot be a tamed and determined matrix of bits, but must rather be understood as the principle of the yet-unthought and unintelligible, of the intrinsic randomness that forever challenges and distorts our matrices, that threatens our models of the world with topological catastrophes and revolutions.

And for these reasons, as we will see, aesthesis can likewise no longer be understood as a mere effect of causation, but must rather be resituated as the condition of causation as such. Though Deleuze develops the concept of *quasi-causality* in his comments on the stoics, it is quite telling that he also reads Leibniz's concept of "compossibility" along similar lines: the reciprocal compossibility and compatibility between factual events selects the world that "will have been." Any shift in paradigm involves aesthesis as a new integration of indistinctions, a new adsorption of chaos, a new inclusion of the hitherto inconceivable. Such speculation finds support in the ubiquitous quantum-physical phenomenon of decoherence. Decoherence is the process by which the superposed states of quantum reality, where everything is "smeared out" in time and space, transitions to the "classical" world of macroscopic objects, where things obtain their specific properties and locations in space and time. With a certain amount of conceptual finessing, it can be aligned with the question of how the physical universe transitions from the "and . . . and . . . and . . ." to the "or . . . or . . . or . . . ," in Deleuze and Guattari's terminology, or from the "inconsistent multiple" to the "count as one," in Alain Badiou's. It is well known today that decoherence depends on the quantum entanglement of the degrees of freedom of the universe: as they become selectively entangled with the already decohered environment, their future trajectories become increasingly correlated with the environment, hence producing the phenomenon of locality on the macroscopic scale. Thus, in some sense, locality *results* from entanglement. Following Giulio Tononi's theory of consciousness as "integrated information" and its quantum physical construal by Max Tegmark, we can think of this mechanism as resulting from an activity of synthesis. As we will see, this echoes my account of the ungrounding of matter and the coarse-graining of nature as inseparable from aesthesis as a general process. In sum, I will suggest that the universe is *quasi-causally* conditioned by aesthesis to coarse-grain itself, giving rise to causation, locality, the intelligibility of space and time, and correlated forms of subjectivity.

Now, this highly speculative investigation will no doubt fit uncomfortably with many established conceptions of aesthetics, especially those deriving from the discipline's characterization as a critique of *judgments of taste*. But fundamentally, aesthetics is not about beauty. It is about perception, experience, and subjectivity; it is about how it is that the "first-person" perspective fits into the broader picture. If, in the world "out there," beyond subjectivity, there seems to be no distinction between beautiful and ugly, good and bad, this and that, here and there, or then and now, then aesthetics is ultimately the question of how exactly these distinctions come into being. I have not attempted to engage directly with all of my contemporaries who are taking positions on this terrain, but my hope is that this book will serve as a contribution to current discussions of these topics.

Handwritten annotations (top):

Baumgarten: aesthetics as a cognitive faculty different from that which allows us to distinguish resolve the distinct details of the object of knowledge

1 CHAOS AND COGNITION

Handwritten annotations:

Leibniz: two aspects characterising the problematic of apprehending the object of knowledge

[capacity of agent doing the knowing]

'clear': a positive subjective capacity ←   the dim or obscure pole characteristics a subjective incapacity to apprehend the object

① a spectrum spanning from obscure to clear
→
② a sp dimension spanning from confused to distinct

Aesthetics and the Indistinct

Aesthetics has always concerned an economy of distinction and indistinction. Alexander Gottlieb Baumgarten, who coined the word "aesthetics" as we use it today, drew his inspiration from the description by Gottfried Wilhelm Leibniz (and in Christian Wolff's derivative Leibnizian philosophy) of a cognitive faculty different from that which allows us to resolve the distinct details of the object of knowledge. Leibniz recognized two aspects characterizing the problematic of apprehending the object of knowledge that are relevant to this concern: the first specified a spectrum spanning from obscure to clear, and the second a dimension spanning from confused to distinct. Inheriting these concepts from Descartes, who had popularized them, Leibniz was more specific about how the two aspects differed.

On the first spectrum, clear–obscure, the dim or obscure pole characterizes a subjective incapacity to apprehend the object. The obscure is that which is "insufficient for recognizing the thing represented," a failure to remember "sufficiently well to recognize," an "inability to distinguish," or an inadequate explanation. In sum, obscurity depends on a deficiency on the part of the knower, not on the intrinsic character of the object being apprehended. At the other end of the spectrum, therefore, we find a positive subjective capacity: "knowledge is *clear* when I have the means for recognizing the thing represented."[1] If I know something clearly, as a subject, as an agential knower, I have succeeded in applying the correct mode of approach and have thus gained perfect insight into the object. In this way, the first spectrum concerns the capacities of the agent *doing* the knowing. It is thus a dimension of variation that is completely

wave dynamics :- continuous spaces- there is not definite regulation of individual quantities (diffraction)

subordinated to the pregiven relation between subject and object, for it concerns specifically a proficiency or aptitude (or lack thereof) of the subject to unilaterally resolve the object.

But, crucially for our purposes, there is a second aspect to the question of knowledge as described by Leibniz, a second spectrum intersecting cognition. Not only is my knowledge either obscure or clear; my knowledge can also be confused or distinct. Some objects of knowledge, no matter how hard I may try, and even if I am applying the appropriate kind of analysis, will forever remain indistinct. Leibniz was of course acquainted with such objects of knowledge, as he was the originator of calculus, which allows one to approximate the infinitesimal subdivisions of the continuous function. In the continuous spaces describing classical dynamics (objects moving in space, thermodynamic dissipation, magnetic fields, wave dynamics, etc.), one never arrives at a definite black or white resolution of the individual quantities. One has to approximate a "common measure through a continual analysis that consists of dividing first the one then the other."[2] He knew that, since such values could only be approximated, in reality, no matter how many times one applied the calculation, one would still never arrive at a black or white answer. Thus, such knowledge is *intrinsically indistinct*.

INTRA-ACTION

DIFFRAC TION

intrinsically
indistinct

> I cannot enumerate one by one the marks sufficient for differentiating [them] from others.[3]

Leibniz
aligned
Qualia
with
mathematical
infinitesimal

Leibniz was making a very subtle connection: he was aligning what we today refer to as *qualia* with the mathematical infinitesimal, the smallest nonzero quantity: $1/\infty$. Though this was not obvious, he noticed that *qualia* are characterized by the same indistinction that contaminates our mathematical approximation of continuous systems. So, in just the way that mathematically incommensurable ratios between quantities lead to "irrational" results, Leibniz claimed that "in contingent truths . . . there is truth, even if that truth cannot be reduced to the principle of contradiction or necessity through an analysis into identities." This would turn out to be the bedrock of the aesthetic discipline, for among the examples Leibniz offered of clear yet indistinct ideas, he included "colors, smells, taste, and other particular objects of the senses," the features typical of internal phenomenal or conscious experience, now often referred to as *qualia*, which are said to be known most intimately and clearly, and yet most confusedly. The immediate data of sense perception are intrinsically confused because it is impossible to explicitly enumerate the individual

TRUTH

Qualia are characterized by
the same indistinction
that contaminates our
mathematical approximation of continuous systems

The infinite progression into the infinitesimal depth of experience, which could be approximated to arbitrary accuracy, defined as the artist's je ne sais quoi, the ineffability of qualia

attributes that allow one to distinguish them from other sense data. The infinite progression into the infinitesimal depth of experience, which could be approximated to arbitrary accuracy, defined the artist's *je ne sais quoi*, the ineffability of *qualia*:

> Painters and other artists correctly know (*cognosco*) what is done properly and what is done poorly, though they are often unable to explain their judgments and reply to questioning by saying that the things that displease them lack an unknown something.[4]

Now, one apparent inconsistency in Leibniz's reasoning here is that he also thought that, objectively speaking, in the eye of God, everything was made of distinct "monads," individual perspectives on the world, little perceptions. How to interpret this caveat has been the subject of much debate in Leibnizian scholarship. According to some, when Leibniz writes that, with regard to *qualia*, "the thing does indeed have such marks and prerequisites into which its notion can be resolved,"[5] he means to say something to the effect that, eventually, with the advancement of technology and conceptual knowledge, we might in fact be able to enumerate all the marks sufficient for knowing it distinctly. Nicholas Jolly argues that Leibniz is "not entitled to say that this is more than a contingent limitation relative to the current state of [science] in his day."[6] Other commentators have claimed that, actually, Leibniz is saying that, no matter how far technology and science progress, we will *never* be able to discriminate distinctly between perceptions, for they are *inherently* indistinct, or at least existentially or *nominally* so. Leibniz's nominalism may therefore imply that the clear yet indistinct knowledge is not absolutely or essentially observer-independently confused, but necessarily *nominally confused*.[7] He certainly did believe that all being was eventually made up of absolutely singular entities, or simple substances.

nominally confused

In other words, though perhaps perceptions *in essence* "have such marks and prerequisites into which [their] notion can be resolved," Leibniz reserves the capacity for resolving these to God alone. For Leibniz, only God can holistically apprehend the indistinct object in its infinite regress of determinations. He recognizes that irrational numbers or "incommensurable" ratios are not "expressible": no finite series of operations will arrive at their complete resolution, even though, in essence and in the mind of God, the marks for distinguishing them are completely seen. Similarly, necessary truths can be demonstrated with "geometrical rigor," as their analysis "arrives at an equation that is an identity." With contingent truths, on the other hand, even though "one continues the analysis to infinity," giving reason after reason, one never arrives at "a complete demonstration, though there is always, underneath, a reason for the truth."[8] Some truths, contingent ones, are irreducible to a specific number or a final determination. As Leibniz claims, even God will not arrive at a final resolution of the contingent truth, for it does not exist as such.

> In contingent truths, . . . the resolution proceeds to infinity, *God alone seeing, not the end of the resolution, of course, which does not exist,* but the connection of the terms or the containment of the predicate in the subject, since he sees whatever is in the series.[9]

Leibniz thus sees the effort of science and knowledge as an *asymptotic* progression toward perfection: though we might approach the complete resolution, there is a gap with the real that subjective perception or knowledge, by nature, cannot bridge. Some objects of knowledge are *intrinsically* confused or vague, even though perfectly clear. Clear knowledge "is either confused or distinct," for though we may have every means possible to see clearly and fully, even with the best explanation or the utmost capacity to distinguish, some knowledge will remain indistinct and confused. It is therefore helpful, with regard to the study of aesthesia, to be careful not to confuse these two dimensions of knowledge (clear–obscure and indistinct–distinct) and their respective orthogonal orientations.

The phenomenon of cinema serves as an illustrative example of how a subjectively experienced perception can nevertheless be subject-independent. We see a continuous motion of the image on screen, even though the image is composed of several still, discrete, and in this sense, *distinct* images in rapid succession. But the *distinction* of these images, on its own, is not sufficient for reproducing the fluidly moving image we see when we watch the film. Leibniz in fact used a very similar example: the transparency we observe on the edges of a rapidly spinning cogwheel. It is the *indistinction* that produces the image we see as we experience a film. Thus, the property of confusion is essential to the experience itself; the distinct images, those *little perceptions,* need to synthesize and integrate into a unified perception in order for *that* experience to occur. Thus *the experience is intrinsically indistinct,* because in order to be *that* experience, it must maintain its characteristic synthetic unity, which depends on the confusion of its parts. An idea or perception that is *of* or *about* an indistinct object will necessarily be impossible to reduce to the constituent

The problematic, the experiential, is the boundary of knowledge, and is therefore also the aesthetic condition of cognition.

parts without radically denaturing its object, or in other words, without committing what is sometimes called a "category error."

An indistinct perception is therefore, in some sense, *real*: it exists independently of the agent's capacities in the specific sense that *no effort of cognition will render it distinct without changing the character of the perception in question.* There is therefore no sense in claiming that *aestheta* are inferior to *noeta*; they simply occupy the boundaries of knowledge, the zones of indistinction. Those aspects of the world that escape the countability of individuality reveal the aesthetic underpinnings of cognition, tied as it is to the organism's encounter with its intrinsically confused surroundings, from which it teases out the provisional invariances and symmetries of its model of objectivity. But, beyond the means or capacities of the knower, the object known is sometimes intrinsically indistinct. Gilles Deleuze pushed this aspect of the Leibnizian doctrine even further by arguing for the reality and observer-independence of the problematic as such. A problem, for Deleuze, is not an effect of subjectivity's incapacity, but rather a real "state of the world, a dimension of the system, and even its horizon or its home: it designates precisely the objectivity of Ideas, the reality of the virtual."[10] For him, therefore, "it is an error [a category error, we might say] to see *problems* as indicative of a provisional and subjective state, through which our own knowledge must pass by virtue of its empirical limitations." The problematic, the experiential, is the boundary of knowledge, and is therefore also the *aesthetic* condition of cognition.

Though philosophy had since the Greeks tackled the general philosophical idea of aesthetics in one way or another, it is telling that its first formalization as a separate category of thought came from Baumgarten, who repurposed the term "aesthetics," which had hitherto referred generally to sensation, to now refer to the *cognition of intrinsically confused knowledge*. Already, Baumgarten gave this study the character of a philosophy of morals that guided the artist or critic in how to properly judge a work of art or a natural object of beauty according to its potential "perfection." In my view, this was an unfortunate downgrading of what the theory could have been held to imply, since it limited its potential by articulating aesthetics on the activity of value judgment. As Kant noted, Baumgarten "hoped to bring our critical judging of the beautiful under rational principles, and to raise the rules for such judging to the level of a science." Importantly, however, Kant rejected the possibility of doing this. "That endeavour is futile" he insisted, because the "rules or criteria" for aesthetics are "merely empirical," and can thus "never serve as determinate

aestheta: sensation
noeta: thought

a priori laws to which our judgmental taste would have to conform." Kant acknowledged that the ancient Greeks already distinguished two kinds of cognition: *aestheta* and *noeta,* or sensation and thought. But the problem for Kant, and the mistake he claimed both Leibniz and Baumgarten made, was the assumption that *aestheta* were actually real things in the world, independent of the observer or experiencer. "Leibniz took appearances to be things in themselves, and hence to be *intelligible,* i.e., objects of pure understanding, (although he assigned them the name of phenomena, because their presentations are confused); and thus his principle of the *indistinguishable (principium identitatis indiscernibilium)* could indeed not be disputed."[11] In essence, what Kant refuted was the Leibnizian proposal that an appearance could ever be "the presentation of the thing in itself."[12] That may have been the fatal end of the idea of aesthetics as the study of indistinct knowledge, preparing the ground for a modern theory of aesthetics as the evaluation of *judgments of taste.* For Kant, it is "our judgment of taste which constitutes the proper touchstone for the correctness of [aesthetic] rules or criteria."[13] He went on to suggest that such a use of the term "aesthetics" should be abandoned. Baumgarten's term did catch on, of course, but the discipline of aesthetics never succeeded in declaring itself a full-fledged science, having been somewhat dogmatically demoted as a lesser science, a lower discipline, or a trivial, even futile, endeavor.

As I have already suggested in passing, in the language of contemporary cognitive science and philosophy of mind, this indistinction characteristic of *aestheta,* versus the distinction of *noeta,* mirrors the problem of the *ineffability* of *qualia.* Thomas Metzinger notes that "we do not possess introspective identity criteria for many of the simplest states of consciousness."[14] This is demonstrated in tests done on the human capacity to identify "just noticeably different" shades of color in the spectrum. Though we are able to distinguish sensuously between one shade and another, we cannot name them or discursively ascribe their difference to a nominal mark. When one is asked to discriminate between two just noticeably different color nuances or subtle nuances of touch, taste, or sonic pitch, words fail to be accurately linked to the faintest sensations. Sensation is always composed of *ineffable* differences. However, whereas Baumgarten and Leibniz would consider these ineffable differences as *intrinsically* indistinct, Metzinger argues that these sensuous experiences are not *sufficient* for knowledge, a perspective that echoes the Kantian account:

You can see and experience the difference between Green No. 24 and Green No. 25 if you see both at the same time, but you are unable consciously to represent the sameness of Green No. 25 over time. Of course, it may appear to you to be the same shade of Green No. 25, but the subjective experience of certainty going along with this introspective belief is itself appearance only, not knowledge.[15]

For Metzinger, the ineffability of the just noticeable difference between sensuously felt yet indistinct intensities is relegated to an incapacity, a lack, or a failure of the cognitive activity. It is implied that, if our "level-two" cognitive function were to be of a sufficient "resolution" or "definition," then it would be possible to resolve such noticeable differences discursively, not just sensuously.[16] Echoing Kant in this regard, such a position implicitly rejects the possibility that real things might be intrinsically indistinguishable, indistinct, or undecidable *in themselves.* Such indistinction is similarly consigned to sensation's failure to determine the matter at hand in finite time. Kant thought indistinction was observer-dependent, an effect of our failures as knowers. This is why he scorned Leibniz for believing that he "cognized the intrinsic character of things" through his "abstract formal concepts." By contrast, for Kant, "appearances are objects of sensibility . . . not pure but merely empirical."[17]

[handwritten margin note: "I don't agree"]

In an analogous way, Kant did not even believe that irrational numbers were really numbers.[18] Because they had to be approximated to an arbitrary degree of precision through an infinite series of operations, he thought, irrationals could not be admitted as objects or concepts of understanding. For him, a concept had to be held within sensuous intuition; it needed to obtain an empirical finitude in space and time. And, since we cannot hold an infinite series of operations within the full presence of our mind's eye, such irrationals, he thought, could not be concepts: "Thoughts without content are empty, intuitions without concepts are blind." We might say that Baumgarten's attempt to formalize aesthetics as a scientific discipline concerned with such indistinct contingent experiences was therefore rejected by Kant for the same reasons. Subsequent conceptions of mathematics did of course make it formally possible to admit irrational numbers as numerical quantities, by considering them in terms of abstract sets (Georg Cantor) or as constructions (L. E. J. Brouwer). It is thus tempting to speculate as to whether aesthetic ineffability could not be conceived in analogous ways. Is the indistinction, the *je ne sais quoi,* of a specific hue of green not itself *constructed* through an infinite, nonterminating series of operations instantiated in the organism experiencing it, a series that never comes to a halt?

Be that as it may, Kant was misrepresenting Leibniz's views somewhat. For, as we have seen, Leibniz explicitly acknowledged that certain kinds of perplexity were caused by our own cognitive failures. For such situations, he reserved the concept of "obscurity." Knowledge was described as an illumination of its object, and its failures left the world in the dark. He recognized the importance of the cognitive agent's capacity to think, to explain, to recall, and to perceive. Leibniz, therefore, did not uncritically attribute all vagueness to the *Ding an sich.* Rather, to reiterate, Leibniz's position was that the dimension from obscure to clear concerned such subjectively dependent aspects of knowledge but the contrast between distinct and indistinct concerned the properties of things in themselves. The two spectra were held to be individually important aspects of the cognitive process. This respect for the formal limits of distinction was typical of such a mathematical mind as Leibniz, who had extensive practice with the symbolic construction of models of the physical world, and particularly the attribution of functions to continuous motion. While Kant rejected the independence of irrational numbers, Leibniz's intimate engagement with the continuum had him embrace irrational entities as real things. Rational numbers, like necessary truths, can be noted, marked, and distinguished with complete precision, but between two consecutive rationals there are gaps of pure irrationality—contingent truths—that are, as Leibniz saw them, the real local compromises of relational entities, constructed asymptotically through a nonterminating series of operations.

There is an ironic sense in which what Kant rejected in the aesthetics of Leibniz, who was the quintessential *rationalist,* was what we would eventually come to understand as a form of "radical empiricism" (to borrow the term of William James) that acknowledged the reality of the indeterminate, the undecidable, the irresolvable, and in some sense suggested that this dimension of indistinct knowledge was inherently tied to the most intimate experience of *aestheta*: colors, smells, tastes, and so on. It is the idea that such *qualia* are both emergent (contingent) and real. Leibniz had effectively likened what we might call "internal states" of consciousness to "observer-independent" states of the world by aligning *qualia* with mathematical indistinction; we might call this Leibniz's *qualia-infinitesimal isomorphism. Qualia* are indeed *petites perceptions.* This was a highly innovative idea that, with Baumgarten, eventually became the implicit

AN EXPERIENCE IS A SYNTHESIS
INTRA – ACTION

foundation of aesthetics as the discipline of assessing the indistinct. At the heart of this idea was the notion of the innumerability or discursive ineffability of the marks for distinguishing one experience from another. In his opposition to Hobbes, for instance, Leibniz rejected the idea that experience or perception could be explained mechanically because he thought that matter is infinitely divisible, while a perception is a unified whole that simply cannot be divided in any way; if it could be, it would simply give way to a qualitatively different perception. An experience is a synthesis, an integration irreducible to the parts that nevertheless compose it. An experience could not be an aggregation, he thought; it needed to be a true unity, and since Leibniz defined matter or extension by its infinite divisibility, it could therefore never fully account for experience. However, Leibniz was nevertheless not supporting Descartes' dualism, the idea that there are two substances, mind and matter. In fact Leibniz opposed this idea just as vigorously, arguing that experiences had to be identical with mathematically indivisible singularities without extension, reflections of the entire world to varying degrees of clarity. Flipping materialism on its head, for Leibniz, the aggregate hinges on its holistic constitution, not merely on the plurality of its parts. The unity of the parts in the provisional whole is somehow what maintains causal priority over the parts it comprises. Thus, these "simple substances," these "little perceptions," these mathematical perspectives on the world, were the basic elements composing all that there is, for the world is nowhere but in its compossible perspectives on itself. Even human experiences, as intimate as they may seem to the human subject, somehow are identical with these extrinsically existing simple substances in their integration. By extension, he saw the relationships of obscurity and clarity between these simple substances as giving rise to the world we experience through their interferences and reciprocal constraints.

Conflict or Compromise

Part of the story of why *aestheta* have traditionally been demoted in relation to *noeta* may be traced back to the origin of philosophy. The pre-Socratics, notably Parmenides, founded philosophy as an exercise in the mistrust of experience. The history of knowledge reads as a progressive questioning of previous assumptions about reality. From the depths of our organismic origins, evolution has committed us to an unexamined naïve realism: as organisms, we have to believe that this event follows that one;

we have to trace effects to their causes in order to survive for any length of time within the environment, in order to escape our predators and obtain means of sustenance. But the philosophical attitude and the scientific reason that is its extension derive from a critique of just those evolutionarily conditioned assumptions about reality. The pre-Socratic explosion of philosophy was inaugurated by the question of being, and responses to this question already insisted that being was something beyond appearances. Right from the beginning, the project of knowledge inaugurated a distrust in what nature had offered to our common sense and a belief that something else resided behind appearances. In its original form, philosophy pitted reflective rigor against the *doxa,* the unexamined opinions of politicians, poets, and artists.

Doxa

But this origin, notwithstanding its venerable subversiveness, may have also inaugurated a prejudice in philosophy that posits two levels of cognition forming a hierarchy in which a clearer and crisper level is preferred over another level that is cloudy and murky. Reason and knowledge came to be understood in terms of dominating or conquering the confusions of the world. Sensation differed from knowledge in that it resulted from the reverse, the domination of the soul by the passions. Anaxagoras thought that, because of the "feebleness [of the senses], we are unable to determine the truth." Most notably, it was Plato's arguments against the "arts of imitation" that inaugurated the iconoclastic theme of subsequent philosophy: the tenth book of *The Republic* sees Socrates convincing Glaucon that poetry, like painting and drawing, "indulges the irrational nature" and "waters the passions instead of drying them up," defining the suppression of the affects as the purest of virtues.[19] A militarist and (as Derrida would eventually demonstrate) *phallogocentric* theme dominates the interpretation: the warrior must be patient and overcome suffering, rather than indulge his affects, in order to attend to the serious matter of assuring victory in the conflict. The conflict comes first. The poets deviously try to make us forget this: they relativize the situation, demonstrating the absurdity and pointlessness of it all. But their deception will be punished ten times over in the afterlife. The progress of knowledge as an unmasking of illusions was thus understood from then on as a kind of confrontation, a war between good and evil, us and them, apathy and emotion, thought and nature, philosophy and poetry, and so on.

binary thinking

This ancient prejudice codified the relation between organism and world as a conflict: to feel was to be won over by chaos, while to know was to dominate it and to impose upon it the order of thought. Ralph

to know = domination

Cudworth reported this explicitly playing out in Plotinus, where "to suffer and to be conquered [are identified], as [are] to know and to conquer": "[Sensation] suffers from external objects [and] lies as it were prostrate under them, and is overcome by them. . . . Sense is therefore a certain kind of drowsy and somnolent perception." Sensations are "confused, indistinct, turbid, and encumbered cogitations very different from the energies of the noetical part, . . . which are free, clear, serene, satisfactory, and awakened cogitations."[20]

The same prejudice was being perpetuated when Democritus claimed that "sweet and bitter, hot and cold, [and] color" were mere conventions and that "in truth there are [only] atoms and the void." It was echoed when Galileo formalized the scientific project, claiming that, "if ears, tongues, and noses be taken away, the number, shape, and motion of bodies would remain, but not their tastes, sounds, and odors," and again when John Locke proposed to distinguish between *primary and secondary qualities*: primary qualities were the object's location and motion in space and the geometries describing its form, while secondary qualities were produced through the subject's encounter with the primary qualities of the object. Secondary qualities did not exist outside the subject–object relation. It is as though the history of ideas in the West was scarred by its origin story, by the decision that inaugurated its essential subversive motif, and never thereafter ceased to echo the idea that to cognize was to dominate the world, like a trauma victim in a cycle of repetitive compulsion. It still echoed, for instance, in Hegel:

> What human beings strive for in general is cognition of the world; we strive to appropriate it and to conquer it. To this end the reality of the world must be crushed as it were; i.e. it must be made ideal.[21]

This same prejudice is repeated again in the deeply entrenched philosophical assumption that there is a strict order of priority between analytic and synthetic reason, or between necessary and contingent truths. Analyticity was supposed to offer insights into the necessary, thus allowing for knowledge to dominate the real, while synthetic or a posteriori knowledge was held to reveal only contingent, trivial empirical facts. It was assumed that "stupid" perceptions, mere sensations, offer no insight into the world, but only report the superficially empirical facts that veil necessary truths. This ancient prejudice, which now finds itself at the crux of our contemporary dismissal of aesthetics as a "lower" discipline, deserves to be critically reassessed. As we are beginning to see, the strict distinction between sensation and knowledge is very difficult to maintain. In a sense, we have come full circle: though we thought for centuries that we were pulling ourselves up by the bootstraps and leaping out of appearances into knowledge, the internal problems encountered by philosophy, mathematics, and science suggest that the history of ideas should rather be read as a progressive expanding of the constraints on the intelligible through experimentation, by tweaking the parameters of sensation and gradually finding real limits and symmetries that order the world, although these turn out to always already be the contingent results of the paths we have taken to get there.

When the famous American physicist Richard Feynman said, "what I cannot create, I do not know," he was touching on a deep truth about how sensation and knowledge are entangled. *It is impossible to formally distinguish knowledge from the sensation of its anticipated construction.* This echoes Henri Poincaré's groundbreaking insight that the fundamental axioms of Euclid's geometry, such as straight lines, flat planes, dimensionless points, three-dimensional space, relative distances, and so on, derive from the coupling between the human's environment and the human's physiology. Notably, Poincaré redefined the geometric point as the class of muscular sensations that accompany the gesture of pointing to it, as one typically does with one's finger. In other words, the difference between two points owes to the difference in the physiological sensations that would accompany the sequences of muscular contractions required for a body to "construct" the points in this way. "What we perceive as signification," mathematician Bernard Teissier writes, "is in fact a resonance produced by our physiology between our conscious thought and the structure of the world as integrated, in an unconscious manner, by our senses."[22] There is "a very strong relation between the *modes of integration* of the data from our different perceptual systems . . . that permit a unified perception of our environment and the *signification* of the mathematical objects that we use to describe this environment."[23] If, in the nineteenth century, the seemingly unquestionable, *necessary* truths of Euclidean geometry were shown to be merely *contingent* by mathematicians like Carl Friedrich Gauss, Nikolai Lobachevski, János Bolyai, and Bernhard Riemann, all of whom discovered coherent geometries that jettisoned Euclid's axioms, then this implies that geometries are derived from the interference of the human organism's corporeal constitution and the particular environment in which it evolves. In a sense, Galileo was wrong: in the absence of eyes, tongues, noses, and ears, *even* numbers, shapes, and motions

all knowledge emerge from organic mixtures.

would disappear, since their preconditional geometries—the spaces in which they are projected, the ordering principles that allow for one element to follow logically from another, and so on—also derive from the organism's encounters with the world, from the practices and the labor through which these encounters are actualized. Deleuze reflected this when he commented that the emergence of language was "a question of a dynamic genesis which leads directly from states of affairs to events, from mixtures to pure lines, *from depth to the production of surfaces.*"[24] But indeed all thought, even the purest a priori and necessary knowledge we may hold in our mind's eye, emerges from the depth of these organic mixtures.

In Solomon Maimon's critique of Kant, there was already an astonishing foreshadowing of such ideas. Maimon argued that we needed not only to identify the *conditions* of cognition, as in Kant, but also to understand its coming into being, its genesis. Kant had tried to secure the synthetic a priori on the grounds that the common-sense conception of space and time is intuited "purely" and is intrinsically indubitable. But Maimon rejects this, arguing that, even though we may find examples in mathematics and the natural sciences that seem to be beyond doubt, propositions that we have never observed otherwise and are unable to imagine otherwise, and thus that seem to be absolutely necessary, this nevertheless does not imply that such structures contain an *objective* necessity beyond our subjective experience of them. In this he was renewing Hume's suggestion that knowledge is an effect of habit. Knowledge is the result of syntheses, concrescences, and integrations of potentially freefloating disparities, unconnected and independent events. Though it is true that knowledge proceeds by analysis and discretization, it does so only insofar as it may reintegrate that which was revealed by the act of dividing and cutting apart.

Foreshadowing the formalization of non-Euclidean geometry, Maimon argued:

> My judgment that a straight line is the shortest between two points can derive from my having always perceived it thus so that for me subjectively it has become necessary. The proposition has a high degree of probability, but no objective necessity.[25]

Indeed, as the discovery of non-Euclidean geometries would soon reveal to the world, the straight line is not necessarily the shortest path between two points; this apparent necessity depends on the constitutive

genesis of cognition

Hume: knowledge is an effect of habit

Our geometries derive from the geodesics our ancestral organisms have adopted in the evolution of epistemaesthetic cognition. (path of least resistance)

geometry in which the points are projected. Euclid's axioms may have for centuries seemed absolutely indubitable, and yet today we know that they were the effects of deeply entrenched habits by which our organism, like its evolutionary precursors, implicitly and unconsciously constructed its spatiotemporal world model, through what Deleuze would call *passive synthesis*. The axioms defining a geometrical system are a matter of construction and mutual consistency, not of God-given objective necessity. Our geometries derive from the *geodesics* our ancestral organisms have adopted in the evolution of epistemaesthetic cognition. They are, in other words, path-dependent compromises between organism and world, rather than Platonic idealities. These questions are precisely those taken up by the Geometry and Cognition research program, uniting contemporary mathematicians such as Teissier, Guiseppe Longo, and Jean Petitot, who interpret the objects of mathematics as *compromises* resulting from such *frictions* between organism and world. We are here quite far from the theory of knowledge as the domination of nature by the organism; for, under the internal pressures of mathematics and philosophy, it has become clear that knowledge can more appropriately be regarded as negotiation and conciliation, rather than conquest or suppression.

This all comes back to Hume: the symmetries and orderings that populate knowledge as a provisional model of objectivity result from a "constant conjunction." Causality itself is a construction of habit. It is the product of an ongoing empirical activity of abstraction. "Beyond the constant *conjunction* of similar objects, and the consequent *inference* from one to the other, we have no notion of necessity of connexion."[26] The abstraction of causal regularity is inseparable from our physiological process of habituation to the world in the creation of our associated objectivity. Like Pavlov's dog, we are *conditioned* to abstract a system of invariances from the chaos around us. As René Thom argued, "we believe in causality because we have been conditioned phylogenetically to do so by the regularity with which phenomena succeed one another in the physical world."[27]

According to the Poincaré-Berthoz isomorphism, for instance, such a constant conjunction is at the root of the isomorphism between the visual geometric line and the "vestibular line" proprioceptively drawn within perception by the inner ear's registration of changes in the body's acceleration. The organism finds a path of least resistance, or a *geodesic*, in the transformation of both systems—visual and proprioceptive—and thereby registers the invariance in which the geometric line is founded.

Teissier argues that the commonplace notions of sequence and continu-ity, as well as the mathematical concepts of ordinality (set theory) and of boundary (topology), descend directly from this organismic identifica-tion between the visually represented line and the proprioceptively felt or anticipated gesture for constructing it.[28] Math, it seems, emerges from habits inscribed in the functional constitution of the organism in relation to its environment, implying that we cannot formally disentangle these apparently necessary truths from our practical and contingent engage-ment with the world.

It is worth recalling here that Maimon's skeptical attitude toward Kant's transcendental aesthetic echoed an earlier, very significant polemic between Leibniz and Newton. Newton thought that space was forever given and unchanging and that time was constantly ticking along at a constant rate. Though this idea certainly did catch on, it is (a version of) Leibniz's counterargument that was eventually vindicated by Einstein's theory of relativity. Leibniz argued thusly:

> I hold space to be merely relative, as time is; . . . I hold it to be an order of coexistences, as time is an order of successions. For space denotes, in terms of possibility, an order of things which exist at the same time, considered as existing together; without enquiring into their manner of existing. And when many things are seen together, one perceives that order of things among themselves.[29]

General relativity posits that there is no pregiven space, no preexisting time: both space and time are effects of the relative masses and velocities of the bodies occupying *spacetime*. Contra Newton, Leibniz had already envisioned such a radical relativity and argued that there is no such thing as absolute space and time. The unilateral determination of the particular by the universal, on which the Kantian system depends, is again put into question by this relativity of space and time. Indeed it casts doubt on the central tenet of Kant's *transcendental aesthetic*. Einstein himself declared: "one cannot stick any longer to Kant's system of a priori concepts and norms."[30] Furthermore, Kant's *transcendental analytic* is also refuted by the radical uncertainty found in fundamental physics following the dis-covery of the strange behavior of the quantum world.

It is moreover interesting to note the contemporaneity of these physical-science refutations of the synthetic a priori and the challenges also faced by analytic philosophy. Moritz Schlick's influential claim that "there are no synthetic judgments *a priori*" (a rejection supported by Einstein[31]) had cleared the way for Rudolf Carnap to argue that we could therefore identify the a priori with the analytic, and the a posteriori with the synthetic. This suggested that philosophy could now implement "Hume's fork" and adjudicate between propositions that are verifiable as true or false and those that are unverifiable and, thus, amount to pointless speculation, thereby restricting the scientific project to cogent and worthy objectives. Willard Van Orman Quine's argument against *logical empiricism* reveals the futility of this prospect. It demonstrates that the priority given to analytic propositions by the logical positivists was unfounded: analytic statements surreptitiously smuggle in their verifiability through an implicit operation of self-justification. This effectively means that the distinction between metaphysics and science is ultimately undecidable, that the two are entangled.

This result would lead Richard Rorty to argue that, if the distinction between science and metaphysical speculation is so vague, fictions and works of art may be just as cognitively valuable as "verifiable" empirical experiments and "falsifiable" scientific theories. In this, Rorty converges with continental philosophy in a rejection of the false priority given to science over aesthetics. The implications of Deleuze's transcendental empiricism indeed come very close to Rorty's ironist pragmatism in which speculation, imagination, and aesthetics are on par with rational thought: science, philosophy, and art are each "creative" ventures, though they result in different products.[32]

Be that as it may, some contemporary philosophers of science still support a loosely Kantian model, though with important modifications and rehabilitations. These adjustments, to some degree, align the neo-Kantian tradition with the principle of the entanglement between distinct and indistinct aspects of cognition. In the face of relativity and quantum mechanics, which seemed to vindicate Leibniz and Maimon against Newton and Kant, the neo-Kantians argued that what was needed to salvage Kant was that we relativize or restrict the synthetic a priori. This neo-Kantian relativization of the a priori originates in Hans Reichenbach's rehabilitation of Kant after the contrary empirical revelations of Einstein's theories of relativity. Reichenbach notes that, in Kant, there were two meanings of "a priori": "It means 'necessarily true' or 'true for all times,'" and secondly, 'constituting the concept of object." But Reichenbach realized that Einstein's theories, as well as the non-Euclidean geometries that paved the way to the theories, put into question the first sense of a priori: there are no stable a priori conditions for all cognition, at all times, in all circumstances; they are somehow relative to the act of observation, dependent on the experience itself, rather than independent of it. Thus, he concluded, "'a priori' means '*before* knowledge,' but not 'for all time' and not 'independent of experience.'"[33]

It is only in this light that Kantians like Michel Bitbol can today claim that something like the transcendental aesthetic still holds for the everyday human environment, given that the Newtonian view of classical physics provides a fair approximation of physics for ordinary human observers.[34] Reichenbach similarly suggested that even the theory of the "flat earth," displaced in the fifteenth century by the spread of knowledge of its spherical character, still effectively functions in everyday practice; at our scale, at our common level of description, we do not have to consider the earth's weak curvature to, for instance, construct a house. This is because the constitutive principles of cognition "do not say *what* is known in the individual case, but *how* knowledge is obtained; they define the knowable and say what knowledge means in its logical sense."[35] Thus they are like a program: the a prioris are the *types* that specify the *construction* of each *token* of experience or *element* of knowledge.

This can be likened to what Maimon was already proposing, that a prioris are differential, or always relative to the manner in which the concept unifies the manifold, the modality of the construction used in each case of cognition: sensation "provides the differentials" from which "imagination produces a finite determined object."; "these differentials of objects are the so-called *noumena*; but the objects themselves arising from them are the *phenomena*."[36] Thus, Maimon argues that the understanding's business of "producing unity in the manifold" is a *constructive* process and is subject to the relativism of the manner in which its differentials are synthesized, the mode of its actualization. The a priori is not absolute, but relative to the process: knowledge is always conditioned from apparent necessities that seem to emerge spontaneously from the recesses of pure thought, but these are different on each occasion, relative to each event of cognition, which is always a compromise between organism and world. Thus, the geometries, symmetries, and orderings that make up the metastable intelligibility and relative predictability of our world neither *represent* the world nor *dominate* nature, but are progressively secreted in the genesis of the organism–world system. The distinct thought or object of knowledge emerges from the compromises of those cloudy and indistinct cogitations that populate the greater sensorium, not as their conquering, but as their conciliation.

Aesthesis Beyond Foundation

The path to the *thing in itself* is paved with an infinite number of deferrals. Like Zeno's arrow, we never reach the target: with each step, the ground quickens deeper into the subtext, and we never gain an inch. The chain of justifications for any specific epistemic or propositional content is never complete, never comes to a halt, never finds a terminus. This was playfully demonstrated in Lewis Carroll's famous story "What the Tortoise Said to Achilles."[37] It shows that the fundamental implicative notion of *modus ponens*—if *p* implies *q*, and *p* is true, therefore *q* must also be true—is haunted by a strange paradox, an infinite regress of conditioning justifications. In Carroll's allegory, modeled on Zeno's paradox, the tortoise tricks Achilles into revealing the ungrounded cascade of regress from which truth conditions emerge. He asks Achilles to convince him that proposition *z* is true given two conditionals (*a* and *b*) that seem to logically imply that it is.

what Joyce does - language differentials [handwritten margin note]

> *a*: Things that are equal to the same are equal to each other.
> *b*: The two sides of this Triangle are things that are equal to the same.
> *z*: The two sides of this Triangle are equal to each other.

The tortoise accepts *a* and *b* to be true, but then claims to be unconvinced that these necessarily imply *z*. Any hypothetical proposition can be judged in terms of its truthfulness, and so, in all rigor, the statement "if *a* and *b* are true, therefore *z* is true" can itself either be accepted as true or rejected as untrue. Carroll shows that this proposition can hence be included as an additional condition of *z*, resulting in the following:

> *a*: Things that are equal to the same are equal to each other.
> *b*: The two sides of this Triangle are things that are equal to the same.
> *c*: If *a* and *b* are true, *z* must be true.
> *z*: The two sides of this Triangle are equal to each other.

But now the Tortoise can again claim to be unconvinced about *z*, for it would require a further claim to be true, namely that:

> *d*: If *a* and *b* and *c* are true, therefore *z* is true.

And of course this proliferation of conditionals can continue indefinitely. It is a paradox of regress that shows that assenting to a hypothetical proposition, and thus halting the regress, involves a *leap of faith* in which we blindly accept the premises to be true, in effect leaving the domain of pure propositional implication. Deleuze writes: "When we say 'therefore,' when we consider a proposition as concluded, we make it the object of an assertion. We set aside the premises and affirm it for itself, independently. We relate it to the state of affairs which it denotes, independently of the implications which constitute the signification." "The conclusion can be detached from the premises, but only on the condition that one always adds other premises from which alone the conclusion is not detachable." In other words, "implication never succeeds in grounding denotation, except by giving itself a ready-made denotation, once in the premises and again in the conclusion."[38] In the proposition's articulation on the "therefore," the proposition cunningly smuggles in an unfounded truth condition, in the blink of an eye.

Our inferences are swept up in an endless cycle between condition and conditioned, each perpetually referring back to the other. Signs are ungrounded, forever referring to each other in an endless retreat: they "can never exercise [their] role of last foundation." No matter how we define the proposition's truth condition or "form of possibility," whether it be "logical, geometric, algebraic, physical, syntactic, [it] is an odd procedure, since it involves rising from the conditioned to the condition, in order to think of the condition as the simple possibility of the conditioned."[39] In this way, the proposition's designation or indication of a state of affairs "remains external to the order which conditions it, and the true and the false remain indifferent to the principle which determines the possibility of one, by allowing it to subsist in its former relation to the other." The failure arises from overlooking the paradoxical material element from which the truth condition emerges. Deleuze writes: "For the condition of truth to avoid this defect, it ought to have *something unconditioned* capable of assuring a real genesis of denotation and of the other dimensions of the proposition. Thus the condition of truth would be defined no longer as a form of conceptual possibility, but as ideational material or 'stratum,' that is to say, no longer as signification, but rather as sense."[40] It would be a matter of integrating paradox within logic itself, of understanding the intelligible as a provisional crystallization of chaos, hardening under the combined pressures of the mobile, incorporeal events occurring on the contingent surface of aesthesia.

Deleuze's take on this regress of justifications reflects developments in philosophy concerning the problem of foundationalism. Philosophy in the last century has faced numerous epistemological difficulties centering on the possibility of establishing a *noninferential* or *self-justified* basis for knowledge. Experience, or the given, or the present, or the real—what Bertrand Russell called *knowledge by acquaintance*—always seems to fail as a foundation for what we think we know, what we think we are saying, what we think we mean. In continental philosophy, a similar realization took place with Derrida's notion of trace and its logic of supplementation. Knowledge has revealed itself to be formally irreducible to the simple events of empirical experience: from within the proposition, we cannot reach the outside to which it should seem to refer, but go infinitely from sign to sign, belief to belief, proposition to proposition. The blocks on foundationalist reductionism in philosophy indeed echo the crises of foundations of modern mathematics. Kurt Gödel's incompleteness theorems suggested that "all that is observable is not rationally deducible from elementary principles and facts."[41] There is a gap that cannot be formally bridged between perceptions as the *implicit* conditions of meaning and knowledge as *explicit* demonstrability. As Longo and Francis Bailly put it, "semantics exceeds syntax," for there is no way to recuperate all meaning from logical formalism.[42] It is important to understand how deeply these ideas intersect the purview of aesthetics as a discipline. For, while it had been thought that knowledge was founded in encounters with the "real" through experience, where our concepts and propositions were to found themselves and acquire their stability, it was now being shown that this was not the case, and could not be.

In his groundbreaking critique of the *dogmas of empiricism*, Quine argued against both analyticity and reductionism. Propositions about the world can be "founded" neither in empirical simples nor in logical analyticity. His twofold attack on logical positivism compellingly contended, first, that what seem to be purely analytical statements have no special status with regard to synthetic statements, and thus have no special explanatory power, and second, that no propositions are fully "determined" by any particular experiences or empirical facts, but rather, that a proposition's truth value is determined holistically and relatively, through the reciprocal tensions composing human knowledge as a field of interconnected inferences. Seemingly a priori analytic truths are truths only because they surreptitiously "smuggle in" their truth value in an implicit manner, through synonymic definitions that veil their reference to other concepts. In other words, analyticity is leaky: the a priori is just the part of the a posteriori whole that is not acknowledged as such. Quine's corrective argument for epistemic holism involved thinking of knowledge as a "total field" or a "sphere" that is "underdetermined" by experiences: "There is much latitude of choice as to what statements to re-evaluate in the light of any single contrary experience. . . . No particular experiences are linked with any particular statements . . . except indirectly through considerations of equilibrium affecting the field as a whole."[43] Thus, inferences cannot be neatly reduced to logical formalism any more than they can be abstracted from individual empirical facts. The field of knowledge can be built up neither from logically indubitable axioms nor from any set of verifiable empirical observations.

Wilfrid Sellars can be said to have contributed a related blow to the prospect of foundationalism with his critique of the "myth of the given."[44] He helpfully noted that the very idea that knowledge was founded in the givenness of experiences rested on two suppositions: (1) that the given be epistemically independent, meaning that it not depend on any prior epistemic content, and (2) that the given be epistemically efficacious, that it make some epistemic difference for the knower. The problem, he noticed, is that these criteria are *mutually exclusive*: nothing can respond simultaneously to both of these benchmarks. To be epistemically efficacious, something must have a propositional structure: it must imply or infer something to be the case. But then this means that, in order to be epistemically efficacious, a thing cannot be epistemically independent, for being propositional in structure precludes that possibility, since propositions necessarily rely on auxiliary epistemic contents.

Heidegger and Derrida should also be included in this discussion. For, again, the arguments in *Being and Time* already suggested a retreat, an essential regress of being, and thus of any stable foundations for conceptual thought: being is projective and fleeting; it extends beyond the immediately present, and therefore is absented by the given object of knowledge or perception.[45] Stemming from his observation that Aristotle's understanding of time privileged presence and the assessment that Western philosophy had in turn perpetuated an unwarranted bias favoring what is *present to* human awareness, Heidegger could then critique Husserl's phenomenological striving for *presuppositionlessness* as simply reviving an ancient philosophical misconstrual of being as immediately *given*. Derrida later extended this critique of the *metaphysics of presence*, taking it at least one step further. His practice of revealing the ways in which a theoretical system (or any sensible structure of ideas) can be reduced to the fundamental dichotomy on which it privileges a presence over an excluded, absent other showed that the system could then just as coherently be reconstructed while privileging the opposing term. Deconstruction thus reveals the arbitrary and prejudiced nature of these decisions, and puts into question the entire edifice of philosophical and scientific knowledge.

Derrida thought Heidegger had not been radical enough, finding that an element of the metaphysics of presence still remained in Heidegger's elevating of the voice and simple poetic phrases over the written word, repeating the central dogma of the metaphysics of the original and the imitative (Plato). In order to transcend this prejudice, Derrida thought, we need to perform the critique of presence from within writing itself. Texts refer to and comment on other texts, not the world "out there": there is a primordial distanciation, estrangement, and self-effacement that immediately underlies all tracing, mapping, and tracking of the real, all that is said or written about the given. Such is the dual logic of differing and deferring united in Derrida's notion of *différance*. The traces that characterize writing efface themselves, *differing from* but also *defering to* other traces, constantly supplementing themselves for others, in a field of coimplicating marks, notes, and references that never end up "representing" anything outside the field. This famously led Derrida to provocatively claim that there was nothing "outside" the text.[46] Here it is impossible to avoid comparisons with Donald Davidson's claim that "nothing can count as a reason for holding a belief except another belief"; there seems to be no way out of the endless chain of ungrounded, unfounded deferrals to other inferences.[47] Even the purest encounter with the real is always tainted with prior context, dependent on prior epistemic contents. It seems that a perception or experience is nowhere but in the "proposition which expresses it—whether the proposition is perceptual, or whether it is imaginative, recollective, or representative."[48]

Western philosophy had hoped that knowledge's chains of justifications would be truncated at some point by the sense datum, by the given, by the self-justifying truth, but unfortunately, this assumption revealed itself to be misguided. If knowledge and intelligibility cannot be "founded" in specific experiences or facts, it is because intelligibility itself emerges from an abyss of indistinction. Again, these developments imply an important challenge to Baumgarten's positing of aesthetics as the science of the inferior cognitive faculty. For, if knowledge cannot find its way down to the sensations to which it is meant to refer, if inferences and propositions are untethered from experiences, then what exactly is this sensation or experience that aesthetics is supposed to be the science of? And how is this science supposed to take place? If it is deemed impossible to retrieve anything knowable through direct encounters with experience, aesthetics as a discipline might seem doomed to fail under Wittgenstein's "proposition 7": "Whereof one cannot speak, thereof one must be silent." But this assumption would be misguided. For, quite interestingly, the discipline of aesthetics is actually indebted to the contrary assumption. The idea of aesthetics as an acknowledgement of and concern for the reality of *indistinction* is inseparable from the Leibnizian rejection of Descartes' claim (which in a way foreshadowed proposition 7) that "we should refrain from giving assent to matters which we do not perceive with sufficient distinctness."[49]

Michel Serres makes this plainly clear: if Descartes argues that we should not admit any doubt into our system of knowledge, Leibniz in turn argues that, if we were never in doubt to begin with, we would never acquire any knowledge at all: knowledge is always indebted to speculation. Aesthetics as a discipline was indeed founded on the idea that we should always consider and respect what resists neatly fitting our current categories of discernment, because it is this inclusion itself that allows us to make a first step toward knowledge. For Leibniz, according to Serres, "the *weight of the ideal of invention balances that of the exigency of certitude*; . . . the progressive dynamism balances the retrospective assurance of truth, in sum; . . . the idea of the general advancement of the sciences balances the ideal of stability or of security."[50] If we always waited to be surefooted, we never could take a single step. The problems of aesthetics are inseparable from the question of how it is that experience happens in the first place. How is it that we come to know what we know? How is it that we pass from obscurity to clarity? How does the real in general, as a realm of possibilities, relate to the world as it factually happens to take place before our eyes? What is the genesis of intelligibility? We might say, therefore, that aesthetics has always followed, not the motto of Wittgenstein, but that of Valère Novarina: "Whereof One Cannot Speak Is What Must Be Said." For, without these incursions into indistinction, without *aesthesis*, which is the true name of speculation, the distinct could never have been apprehended.

Intelligibility is bathing in paradox, the product of an ongoing contingent process. And sensation is not merely an access to the real that could *found* the rules of logic and justify our distinct cogitations, but rather the monstrous indistinction that constitutes the boundary of the intelligible.

From the proposition, we can never find our way out of the regress and indefinite proliferation of supplementary signs, traces, and references. But we can understand the very paradox of this proliferation as *the surface of translation* between what the proposition expresses and the attribute of the state of affairs. To reintegrate Deleuze's thought on this, the paradoxical, preindividual depth that envelops and conditions all intelligibility must be reimagined as the very boundary between propositions and things. As on a Möbius strip, the inside and outside are obtained through a smooth translation upon a single surface. There is thus an immanence of thought and being as much as an estrangement. What the proposition expresses, the semantic depth that exceeds its syntactical formulation, is what Deleuze calls "sense." It is that which generally eludes logical syntax but is nevertheless the condition of its intelligibility. As was shown by Gödel's incompleteness theorems, there is always something the system expresses that cannot be recuperated by the system's formalism. "I can always take the meaning of what I say as the object of another proposition, of which, in its turn, I do not say the meaning."[51] But this is no reason to deny the reality of this illogic that subtends logic; in fact, it must be incorporated into any rigorous assessment of aesthesis and cognition.

The reinvention of aesthetics beyond the myth of the given and the metaphysics of presence, I would like to suggest, requires that we shift its focus from a vague understanding of sensation as a *foundation* for knowledge to the Leibnizian idea that experience precisely concerns the regions of indistinction, the paradoxical, undecidable, uncountable, preindividual limits of rational thought that condition the strictures and coherences of intelligibility. Sensation is not the foundation of knowledge, but the paradoxical indistinction on which it rests, and which is always imported implicitly into the truth statement in the operation of the "therefore." Admitting that knowledge can be founded neither in logical atoms nor in individual events of experience, we can consider the aesthetic not as the stable ground on which all knowledge rests, but rather as the indistinction from which objectivity and intelligibility always emerge to provisionally structure the world in terms of patterns and invariances, order the chaotic environment, and provide it with an impermanent metastability, predictability, and reliability. Aesthesis is the process by which *seemingly necessary truths emerge from the compossibility of contingent events.*

This view furthermore requires accepting that the old idea of a hierarchy between higher and lower cognitive faculties was unwarranted. Rather, we must recognize that there is no way to disentangle aesthesis from distinct cognition: we must perhaps go as far as doing away with the strict distinction between *aestheta* and *noeta*, between aesthetics and epistemics, and posit a single hybrid category, an *epistemaesthetics* that strives to model the genesis of the progressive constructions that direct our thoughts, the necessarily provisional architectures of sense that order our concepts and provide the conditions of all intelligibility, as they transition from indistinct to distinct.

Between Seeing and Knowing

Since the Greeks, the concept of knowledge has been pervaded by the idea that sapience is strictly distinct from sentience, that to know is qualitatively different from merely believing, feeling, sensing, or seeing. This of course reinforced the narrative of human exceptionalism: though animals perceive, and can thus, in a vulgar sense, be said to "know" their environment ("knowledge by acquaintance"), they do not "know that they know," as we humans do, for they cannot give reasons or justifications for their beliefs. According to this line of thought, humans are different from animals in that they lift off from mere sentience and dominate their senses, gaining a top-down view of their perceptions. In a formulation originating in Plato, knowledge was thus defined as *justified true belief*. If knowing was different from merely perceiving or believing something to be the case, it was because knowledge was a belief that was actually correct; it represented, *corresponded* to, the world as it stands, independently of the knower. Furthermore this belief needed to be *justified*: one needed to have the belief for the *right reasons*.

   The problem with all this, as was demonstrated by Edmund Gettier in 1963,[52] is that one can never really confirm that one actually has knowledge: one can never *know that one knows*. This is because one can always, in principle, construct a case where one indeed has a justified true belief, but one that is justified for the wrong reasons, such as when correctly telling the time by glancing at a clock that has been stopped for exactly twenty-four hours. Is this knowledge? You tell the time correctly (it is in fact 4:23 p.m.) and you are justified in your belief (you assumed the clock was running), but you were just lucky to have glanced at it exactly twenty-four hours after it stopped. The problem is that, by introducing such hypothetical situations, one can always doubt whether one really knows what one knows, for in verifying one's justifications for believing that such and such is the case, the justification's confirmation of your belief *could* be

just a fluke. One can always oppose a possible "defeater" to the justification, breaking the causal chain between the belief, the justification, and
the truth. The Gettier cases can thus be said to have demolished the wall
that stood between sentience and sapience, because it no longer is clear
that knowledge in any way lifts off from senses, perceptions, or beliefs.
Indeed, our chains of justifications are just complex networks of coimplicating perceptions, events of experience that constrain each other, but
from which we can never disentangle ourselves to gain an exterior perspective. Analogously to constructivistic or intuitionistic interpretations
of mathematics, in which the mathematical object is entangled with the
activities of the organisms constructing them, one of the lessons of the
Gettier cases is that knowledge cannot be fully disentangled from believing and can never claim its strict independence from mere sensation or
perception. Rather than a difference in kind between human sapience and
animal sentience, we can posit only differences in degrees of complexity.
Though we most definitely know things, we do not know *that* we know
these things, because, as it turns out, knowing is a just a sophisticated
way of perceiving.

An analogous realization of this entanglement between perceiving and
knowing also took place within art history. In a sense, Western art came to
this understanding from the opposite direction, through the trials of its
own historical tectonics. It had been thought for centuries that the role
of art was to produce faithful representations of appearances and perceptions. But it is well known that, once the artistry of linear and color
perspective seemed to have been fully mastered, illusionism and representation were almost immediately abandoned as the primary motive
of art. Why is this? The answer eventually provided by the impressionist
movement was that, once we had conquered the geometry of linear perspective (the quest of Italian Renaissance art) and, subsequently, the relative measures of color perception (which can be said to have become the
primary concern of North-European painters), we immediately realized
that the images we were producing still did not capture the way we see
the world. Something was being left out of the picture. Far from producing
faithful representations of appearances, Western art had until now been
forcing appearances to conform to certain epistemic biases.

When children or amateurs try to produce a faithful drawing of an
object before them, they inevitably resort to what they know: a hand has
five fingers, a house has a roof and a chimney, a sky has a sun and clouds.
The mastery of illusionism consisted in overcoming these epistemic

biases by figuring out how the appearances themselves, in the absence of knowledge, present themselves to the eyes. Western art had been guided by this idea: in order to reproduce appearances on canvas, we need to paint not *what we know* of the objects before us, but *what we see*. This reinforced the traditional distinction between science and art: *seeing is not knowing*. We might speculate that the ancient prejudice to view the relation between appearances and knowledge as a conflict had, with the Renaissance, infected the artistic disciplines too, but they fought for the opposing team: where science had to overcome appearances, the arts had to overcome knowledge.

But by the mid-nineteenth century, artists were realizing that, even though, throughout their conquest of illusionism, they had thought that they were resisting this tendency to paint what they knew by painting what they actually saw, now that that this process seemed to have run its full course, it was obvious that they were in fact still imposing upon reality a certain epistemic mode of its expression to the senses. Within perception, the scene is not always dominated by the vanishing point, and the sky is not always blue. Through the scientificity of linear perspective and the practical knowledge of the interactions between hues of color, artists were still imposing upon appearances the biases of an impure, polluted perception, pervaded by the unacknowledged distortions according to which art was surreptitiously forcing the world to appear to the senses.[53]

This is where calls to rediscover the "innocent eye," an influential concept previously coined by John Ruskin in *The Elements of Drawing*,[54] entered the scene. The goal was to finally, once and for all, rid art of its infection by such epistemic biases, and this can be said to have been what the impressionists set out to do. But again, almost as soon as this project had been defined, it began to collapse under its own contradictions. Ernst Gombrich discussed how, after an initial moment of disruption, impressionism came to be accepted as a true innovation, and common people began to realize that they could take a walk in the forest, squint, and indeed observe nature presenting itself to their senses as scintillating splotches of formless color, echoing Cézanne's assertion: "I see in stains."

But this development ended up having the reverse effect as what might have been initially wished for. We had not discovered an innocent eye; we had rather made a much more important discovery. By ushering in this new way of letting appearances present themselves to our senses, impressionism had shown the contingent and interpretive nature of the disclosure of reality: appearances are *necessarily* conditioned by our technology, our science, our knowledge, our practices, as a matter of principle. With every innovation in the history of visual representation, we had found that an aspect of appearances still resisted our capacities; our knowledge always crept into the reproduction of appearances, for as we were now realizing, there is no such thing as simply seeing without knowing.

By the mid-twentieth century, when the Gettier problem was defined, art had long already converged on the same epistemaesthetic realization. It became obvious that our perceptions too, though not arbitrary, were contingent path-dependent compromises between the world, our organism's evolutionary adaptations, our technological affordances, and the cultural history of our knowledge. This was historically manifested by the experimenting that subsequently took place and the explosion of parallel post-impressionist movements that followed from it. Cubism accepted the artificiality of our attempts at representing appearances by laying bare the geometries and formal biases we imposed on appearances, making the methods of perspective, which had hitherto been eclipsed from the scene, show themselves within the representation itself. Futurism (and in some cases cubism as well) can be said to have shown the processual nature of these artificial and technical constitutions of our ways of seeing, attempting to catch the processes themselves in the act of generating appearances. Surrealism embraced the essentially hermeneutic character of perception by demonstrating, under the influence of Freud, that all seeing was also infected by subconscious factors, dreams, drives and desires. Dalí's toying with forms that could be seen as representations of multiple different objects depending on how one looked at them, picking up on various *Gestalt* psychology experiments, again made apparent the interpretive character of all perception. And of course, the various twentieth-century investigations of abstraction, expressionism, and conceptualism all followed suit. Art, like science and mathematics, was gaining a new and rigorous humility with regard to its practice and discipline. Philosophy and art converged in this new humility, for each had found that it was impossible to objectively or absolutely parse out seeing from knowing, to distinguish once and for all between sentience and sapience. It had always been assumed that philosophy, which was concerned with *noeta,* and art, which was concerned with *aestheta,* occupied perfectly distinct categories of concern. But the traditional wall between the categories of epistemics and aesthetics was collapsing, due to parallel realizations within science, philosophy, and art. Indeed, these realizations were suggesting the need

for a new way of addressing the common concerns of seeing and knowing, of aesthetics and epistemics, of art and science.

As I have been suggesting, we are compelled by these mutual realizations from both sides of the old conflict to compromise and address the
common concerns of epistemics and aesthetics *epistemaesthetically*. We
have already had glimpses of what this might look like. If knowledge is
inherently holistic, we can understand sensation as its *limit*, its *boundary
condition*. In the words of William James: "Conceptual systems which neither began nor left off in sensations would be like bridges without piers.
Systems about fact must plunge themselves into sensation as bridges
plunge their piers into the rock."[55] But we have seen that this cannot mean
that knowledge is *founded* in sensation, or that its chains of justification
find a terminus in individual events of experience. James was echoing
similar reflections in Locke: "The simple ideas we receive from sensation
and reflection are the *boundaries* of our thoughts."[56] In other words, an
Umwelt has a horizon, and this limit corresponds to experience. Even
though no specific knowledge is reducible to any specific sense datum,
and even though the senses are by definition unspecifiable, ineffable,
and indistinct in the Leibnizian sense, knowledge nevertheless emerges
from their concrescence, their synthesis, their progressive crystallization
within the organism. This justifies Quine's remark that knowledge is a
"field of force whose boundary conditions are experience": "A conflict
with experience in the periphery occasions readjustments in the interior
of the field."[57]   DIFFRACTION

These speculative considerations offer us the means to rehabilitate
our understanding of the relation between *aesthesis* and distinct cognition. Aesthesis is cognition's interface, not with indubitable truths or self-
justified beliefs, but with the (as of yet) unintelligible, with non-sense.
Aesthesis does not ground knowledge, and yet all of knowledge must be
said to derive from it: it is a process of translation between doings and
beings, events and memory, randomness and causality. In Leibnizian
terms, we must posit that this transformation of chaos into knowledge,
this aesthesis, follows a gradient from indistinct to distinct, rather than
from obscurity to clarity. What Quine referred to as "total science" is amenable to this: an inherently revisable system of symmetries and synony-
mies, translations, provisional invariances, and ordering principles that
populate our fluctuating and prosthetically mediated relationship to the
chaos that surrounds us.

AESTHESIS IS COGNITION'S
INTERFACE WITH THE (AS OF YET)
UNINTELLIGIBLE, CHAOS AND COGNITION · 47
WITH NON-SENSE
↳ JOYCE'S WORK INITS/PROBES THE
DIFFERENTIAL SPACE

If aesthesis is the process, the transition, then *aesthesia* can be understood as the ongoing result of the process: a provisionally integrated concrescence of chaos. It is not clear, from this perspective, that there is any substantial difference between our total science and our aesthesia. Much like Quine's "field of force," the constraints on cognition loosen at the edges. The bonds dissolve into vagueness and are dominated by infinite speeds—this is what, as shorthand, we call "experience"—while near the center, aesthesia hardens and slows under extreme pressures, giving way to what we call "knowledge."

If aligned with the Leibnizian concept of *compossibility*, this field of reciprocal constraints between perceptions could furthermore explain the apparent stability of causality, objectivity, and intelligibility. For, as we will discuss at length in chapter 4, a strange reversal takes place. Instead of being grounded in necessity, as was knowledge in the Leibnizian system, the synthetic or integrative function of aesthesis can be said to be the *precondition of necessity*: what we call necessity is not what grounds contingency, but rather that which is formed under the contingent pressures of compossibility. The core is inhabited by seemingly unquestionable or necessary truths, while the boundaries are contingent experiences in their purest and most radically unpredictable form. This is because, near the center, the pressures of compossibility are most intensely compelling and obliging. Yet, even the apparently necessary truths found at the core are provisional to their eventual revision by *events* occurring on the surface. With Deleuze, therefore, we can understand events as "pure surface effect[s]."[58] Organisms are world-models. They are networks of reciprocally constraining references to relatively stable factors of the often random and unpredictable world they interface with through various probings and observations. And so aesthesis is the process resulting in the provisional clearing of a region of relative causal predictability and logical stability. Novelty enters the sedimented strictures of the total science from the periphery, and its effects ripple and propagate across the matrix. All events are symmetry-breakings and catastrophes: they irreparably change the topology of the *Umwelt*, remapping its connections, functions, and capacities. Aesthesia is thus no more and no less than the *fabric* of the perpetually revisable sphere of knowledge, a field of integration spanning from the boundary to the core, making it impossible to formally disentangle epistemics and aesthetics. Rigorously speaking, therefore, we must speak of a single *epistemaesthetic* category.

Deleuze approaches a similar conception of aesthetics in his commentary on Kant. As Steven Shaviro points out, Deleuze's conversion of Kant "from transcendental idealism to transcendental empiricism . . . excavates and reveals certain hidden potentialities in Kant's own thought": "It turns Kant away from being a thinker of juridical norms, and transforms him instead into a thinker of singularity and difference."[59] Rather than reading the Third Critique as a trivial part of the Kantian system, as do many Kantians, Deleuze reads it as the culmination of the project, where the critique of the faculties finds its most complete expression, indeed where even the a priori is ultimately relativized by the senses. The Third Critique changes the entire scope of the critical project by discovering that "any determinate accord of the faculties under a determining and legislative faculty presupposes the existence and the possibility of a free indeterminate accord." "According to this principle, despite the fact that our faculties differ in nature, they nevertheless have a free and spontaneous accord, which then makes possible their exercise under the chairmanship of one of them according to a law of the interests of reason."[60]

Thus, Kant's encounter with the strange indifference, disinterest, and unconditioned nature of aesthesis, which Deleuze aligns with Kantian "reflective judgment" to distinguish it from determining judgment, is found to be the condition of all determining judgments as such. In other words, for the faculties to divide into logical, algebraic, geometric, practical, moral, and so on (that is, for any single context to be exhibited as having to be judged according to the rules of a specific propositional form of possibility or truth condition), the faculties must first be allowed to freely engage with each other in an indeterminate manner: the interpretive context and its correlative *interest of reason* emerges from this prior interplay.

> It is true that in the *Critique of Judgement* [sic] the imagination does not take on a legislative function on its own account. But it frees itself, so that all the faculties together enter into a free accord. Thus the first two Critiques set out a relationship between the faculties which is determined by one of them; the last Critique uncovers a deeper free and indeterminate accord of the faculties as the condition of the possibility of every determinate relationship.[61]

Hence there is again a sense in which aesthesis and cognition are inextricable from each other. Aesthesis is the process or genesis of the conditions of intelligibility. When we speak of aesthesis, we are referring to the *transition* corresponding to the progressive integration of otherwise free floating elements that must be organized in order for intelligibility to take hold. Aesthesis can be nothing other that the conversion of the unthought into thought, the phase transition between uncognized and cognized, unintelligible and intelligible. And epistemaesthetics concerns the question of how it is that the conditions of knowledge, intelligibility, order, symmetry, invariance, predictability, and pattern come into being in the first place and allow the organism to obtain a provisional stability within an otherwise indistinct, chaotic, and random environment.

Indistinction, Randomness, Art

We are beginning to resolve the inherent relationship between the intrinsic indistinction of sensation and the production of intelligibility's distinctions. Cognition concerns various forms of synthesis and integration, which necessarily presuppose processes of transition between the discrete and continuous, an economy of distinction and indistinction. But there is, furthermore, also a link between what appears purely random and contingent (indistinction) and what appears to us as novel, original, or causally *independent*. Artists know that the true event appears to have no *specific* cause. That which is original can emerge only from the unpredictable, and so it is not surprising that contingency in its many guises intersects so many aspects of art practice. This is precisely why the artwork must be a "monument," as argued by Deleuze and Guattari, why it must stand for itself: the artwork has to emerge from an independent symmetry, a superposition of causal lineages that can be collapsed only *as* spontaneous, as causally independent. Though it can be said that all practical disciplines are creative, art is a privileged engagement with the precise problematic of the event; artists are called on to produce novel compositions, innovative experiences, and new percepts and affects. Hence, the artist works within a polarized orthogonal temporality corresponding to this transition from potential to actual, but also from non-sense to sense, from unintelligible to intelligible. Art is a privileged engagement with these processes. It presents us with the indistinct as such; it inserts itself within the integrative process only to momentarily freeze it and counter its action, avoiding the collapse of the indistinct into the distinct while nevertheless offering it to cognition, making the process of aesthesis explicit to itself. We have seen how the traditional distinction between sensation and knowledge mirrored a distinction between feebleness and strength and how, according to an ancient prejudice in the Western history of ideas, sensation was considered a failure to dominate the world while the noetic act was characterized as its conquering. But the artist knows that it is not so easy to be conquered by that otherness, to let oneself be invaded by the forces of indistinction. To do so has often been the artist's great challenge.

Perhaps nowhere has this been more explicit than in the radical artistic movements of the last century. The dissociation of the subjective act from the outcome of the artistic event was a general trend in the avant-garde's deconstruction of modernist and humanist aesthetic ideals. There was a turn to the inhuman, the irrational, the sublime speeds of the mechanical, and the ecstasies of the random as sources of an inexhaustible spontaneity. These methods were put to the use of distancing the artist's subjective decision-making process or intention from the resulting artistic product. Many different methods were used to arrive at this disconnection. Dada embodied this most explicitly. For Jean Arp, chance and serendipitous encounters were a manner of reconnecting with the irrationality of nature.[62] For Tristan Tzara, the poetic cut-up technique was a method for arriving at a result that diverged from the intentions of its author, a randomization process later taken up by William S. Burroughs and others. Marcel Duchamp used wind and gravity to produce randomness in his creation of forms.

Though it differed in where it chose to source its randomness, surrealism's early infatuation with automatic gestures can be said to have extended this line of reasoning. Under André Breton's psychoanalytic influence, surrealism ventured to circumvent subjective human intentionality by appealing to dreams, repressed desires, slips of the tongue, and uncontrolled gestures and thoughts. (Breton and Dali later resisted, however, against a total embracing of automatism, both claiming that it should be tempered by rational or critical thought.[63]) The *cadavre exquis*, popular with the surrealists, was a strategy for producing nonintentional aesthetic products through blind collaboration, combining the imaginations and intentions of multiple artists, reciprocally neutralizing them to arrive at a hybrid monster of an image. In the *cadavre,* neither one of us is responsible for the resulting work; it is a product of our combined wills and intents, and thus is the contingent resolution of several streams of causality, repotentialized, poised, as it were, through the reciprocal neutralization of many causal paths, their blurring, their scrambling.

With the revival of Dada in the Fluxus movement of the 1960s, the explicit emphasis was on change, events, happenings, and the rejection of authorship. John Cage implemented the I-Ching divination practice as a method of composition. Maciunas's "Fluxkits" were designed to produce contingent events while scrambling their origin or their authorship.

The "event scores" featured in *An Anthology of Chance Operations*, with contributions from most of the artists and composers associated with Fluxus, epitomize the age's taste for indeterminacy as a source of pure creativity.[64] Less explicitly, though just as importantly, the situationists also used randomness generation techniques to scramble society's psychogeographical control topologies: the strategy of "derive" is essentially a *random walk*, a *markov process*. More generally, a countercultural antimodern spirit of improvisation (jazz) and spontaneity exploded onto the scene in the last century, promoting techniques for distributing the act of composing works across multiple agents, in effect blurring the causal chain so that no one agent could be credited with authoritative creative control over the work. For, even though the mechanisms differed, in each of these practices, the goal was similar: make it unpredictable, random, or contingent; appeal to a monstrous symmetry so that the event may occur in its spontaneous break or collapse, expressing the *quasi-causal* determination of the world.

But the question of how to *care* for contingency, how to be receptive to indistinction, hospitable to it, or how to *conspire* with it, has actually been at the heart of aesthetic practices for longer than we tend to think. These themes deeply intersect art-historical discourse on authorship, style, technique, and practice: from the Greek notion of the *daemon* or the muse to *commedia dell'arte*'s improvised styles and romanticism's investment of fatality, long before the nineteenth-century cults of madness and primitivism and the avant-garde's explicit use of divination, chance, and automatic operations.

Intuitively, this makes sense, since the artistic event, like any event, is necessarily unknown before it happens: it must be unpredicted when it occurs; otherwise it wouldn't be much of an event. Duchamp's notion of the *art coefficient*, the idea that the creative act always escapes the artist's will or intention, was an extension of the reasoning that has always resided in appeals to animistic spirits, daemons, subconscious drives, randomness, contingency, and to any indistinct causal determinants: they demand the anonymity of the causal chain resulting in the event. In this sense, they are no different from the toss of a coin or the shuffling of tarot cards. They say: let the cosmos decide what happens; let an anonymous ordering principle express its *divine will* over circumstances. Let the indistinction express itself for itself. In Kant's Third Critique, this inherent relation between artistic aesthetics and contingency resides in his conception of art as that which must appear to be purposeless or to have happened by accident, while nevertheless being the fruit of a laborious process, or that which must seem to be of universal import, while being an act of the individual will.[65]

Of course, this is also reflected in the common association of artistic genius with madness, melancholia, and the divine. The romantic cult of genius was confused with that of madness, for it was implied that the creative genius was not a *rational* subject, at least not in the moment of inspiration. The genius tests the waters of insanity with every creative act, which implies that the creative act is one that escapes or surpasses the will of the individual; the genius perhaps hears many contradictory voices, and his actions result from their interference. Insanity is also unavoidably linked to divine will. The deformed or deranged individual has always been a *pharmakos* in one sense or another, a peripheral figure announcing the onset of the unknowable and *untouchable* outside the orders and norms of culture. Starting with Michelangelo, if the term "creation" was progressively applied to works of art, it is again by virtue of an implicit connection between the artist's contingent creation and the divine: the artist can be said to channel a cosmic will. Again and again the act of creative genius[66] is dissociated from the discursive and explicit will of the rational individual and placed in the service of the expression of a primordial creativity and unpredictability. In this light, art can be seen as the clearest expression of aesthesis in the domain of human practice.

The divine contingency embraced by the creative act is the result of the resolution of tensions between many conflicting inertias and fluxes in the world, different chains of deterministic constraints, which collide and combine and result in whatever happens. The event is always contingent in this sense, always the result of a plethora of different causal lineages scrambled through the state space, resolving tensions from point to point, finally arriving at *this* moment in *this* place to produce *this* contingent event. Contingency is hence inseparable from the notion of unpredictability. The more contingent the event, the more unpredictable the occurrence, the less it seems tied to any single will or cause when it happens: all causal lineages, all chains of justification, seem to converge on the event in a *symmetrical* fashion. If they did not, if the probability of the event was *asymmetrically* stacked, then one would have a way to gauge the probability of the event and it would lose its characteristic unpredictability, and hence lose some of its contingent character. True events are strictly random. The more an event happens without a specific asymmetrically slanted cause, the more it will have been caused *univocally* by all of the wills to power united, or by a symmetrical, undifferentiated whole of causality. It has been referred to in several ways: chaos, the infinite, the divine, the virtual, the monstrous symmetrical realm from which all potential expresses itself most univocally in those happenings that do not select a cause or that choose the whole of causality for themselves. The event is thus always a symmetry-breaking, a collapse of the actual by the potential: from a blind potential emerges a singular occasion, in its absolute irreversibility and irreducibility. Cognition is inseparable from these symmetry-breakings: each event in experience changes the holistic topology of the field of knowledge, of the "world model" or "model of objectivity" that provisionally defines the real. This explains why humans have been tempted to view remarkable events as an expression of "divine" will, for in truly significant collapses of the topology, no worldly will or specific causal chain seems to directly account for it. The significant event seems underdetermined by any specific cause, but seems rather to emerge spontaneously from all causes fused, con-fused together (in the Leibnizian sense of *indistinct*). And artists, it seems, are in the business of making these symmetry breakings cognizable, of making the very process of world-model collapse appear to itself.

How are these symmetry breakings materially and operationally implemented? When there is an equal probability of an event happening or not happening, then its probability distribution can be said to be symmetrical.[67] Take the common die. Each time the die is cast, there is an equal chance that any of the faces will come out on top, and this equal probability is a direct consequence of the object's "platonic" symmetry. When all outcomes are equally probable in the space of *possible* outcomes, then the probability is symmetrically distributed among these possible outcomes. Hence, the more the probabilities are symmetrical, the less predictable is their outcome. This is why chance devices such as dice will implement some sort of geometrical symmetry: the flipping of coins, the spinning of a roulette or choice wheel, and the shuffling of a deck of cards are all activities that depend on this same scheme. In these, the problem of summoning chance is consigned to a material object that implements a symmetrical distribution of probabilities.

There are also other systems for producing randomness that, instead of relying on the symmetrical geometry of an object, make use of algorithmic procedures. Bamana divination involves a modulo-2 addition algorithm, which is an implementation of a kind of *pseudo-random number generator* much like those of computer programs. You start by drawing four sets of random dashes in the sand. Then you pair the dashes up to determine whether each row is odd or even, recording the result with one dash or two. Repeat this four times to give four different symbols, one at the end of each row. Then recursively modulo-2 these results, applying the same folding reduction onto the outputs, alternatively pairing them vertically and horizontally. The process requires several recursive levels of this procedure, for it is really a game of scrambling the original pattern drawn. Much like a cryptographer's cipher, it is a way of taking a known number and deterministically scrambling it with a complicated set of algorithmic procedures intended to ensure that the diviner could not have consciously planned to produce the given outcome. The diviner is basically implementing a way to randomize his own action, to distance himself from his own intention or free will in an explicit scrambling of the initial gesture.

As revealed by the ethnocyberneticist Ron Eglash, the Bamana divination system simulates *deterministic chaos.*[68] This means that the outcome, though completely deterministic and bound to the simple rules of the recursive process, is nevertheless extremely difficult to predict or to retrace to an initial condition. Eglash suggests that the underlying principles of Cantor's theory of transfinite sets, which tamed the notion of infinity while also unleashing the monsters and paradoxes it contained, found precursors in these African divination systems. He also claims that binary arithmetic, which was eventually formalized by Leibniz and ultimately led to the invention of the computers we rely on today, can be traced back through documented historical links between Western geomancy and African fortune telling uses of modulo-2 arithmetic.

The computer is our contemporary oracle. But we often forget that this was very much Leibniz's dream when he first imagined the computer. What he called his "stepped reckoner" was an early mechanical calculator, a precursor to the computer inspired by Pascal's adding machine (the pascaline). In his youth, Leibniz had been influenced by the Catalan mystic Ramon Llull and similarly speculated that all thought could be translated from natural language—which was subject to interpretation—into an absolute mathematical language, a *characteristica universalis,* a universal script that could precisely translate the truths of reason, allowing for the unequivocal calculation of all problems of logic, and by which "every reasoning derivable from notions could be derived from these notions' characters by a way of reckoning."[69] He thought that, if the universal language was precise enough in its particularity, even if complex in its generality, it could be processed by a machine and give a definitive yes/no

answer to even the most abstract moral questions. And he found somewhat of a confirmation for his idea of a universal language in the dualities
of the I-Ching, the ancient Chinese *Book of Changes* (the same later used
as a compositional tool by Cage), which were introduced to him in correspondences with Joachim Bouvet, a Jesuit stationed in Peking. Attributed
to the mythological figure Fu Xi, the ancient hexagrams were thought to
depict time and the universe through conflicting dichotomies (dark–light,
male–female, yin–yang), and Leibniz recognized that they corresponded
to a base-2 counting scheme, much like the system of numbers he was
developing based on the zero and the one: the binary number system. "All
numbers can be expressed in this way by unity and nothing."

Even though he never got around to building a properly binary computer, Leibniz realized that base-2 would be a more effective way of implementing calculations in machines. His philosophy of combination posited
that everything was composed of absolute identities, simple substances,
and so he speculated that, with a sufficiently robust mechanism, one
could dial yes/no questions into a machine and step through to the correct answer. This dream of such an oracular machine can be said to have
been shared by David Hilbert and formalized in his *Entscheidungsproblem* (decision problem). Though this dream was eventually shattered by
the work of Gödel, Alonzo Church, and Alan Turing, it is important to
underline how the history of computing begins with the prospect of building a *mechanical oracle*. Indeed, Leibniz believed that the binary system
held some sort of mystical power or significance. The computer was itself
first imagined as a machine for automating divine revelation (this can be
said to have returned with the advent of AI, which has spawned a kind of
computational messianism popular with the Silicon Valley speculators,
transhumanists, and singularity theorists who await the technological
rapture). After the eventual failure of the formalist program, however, we
now know that the oracle of computing could not answer all questions
with a *yes* or a *no*, that it is actually dominated by undecidability and
indistinction.

The random procedures of divinatory practices are inherently related
to the conceptions of the divine realm: they grant the human community
access to the infinite. The chance artist, the diviner, and the gambler are
three examples of engagements with infinity through randomness. They
reflect the virtual unity of the faces of the die after an infinite number of
throws; the more the die is cast, the more the statistical tendency converges on a symmetrical distribution as the cumulative throws approach

infinity. Each single casting of the die will result in a single face coming out on top, a break in the symmetry of the cubic object. But, after infinite throws, on average, the faces approach an equal distribution, resulting from the equal probability of any face coming out on top in each instance. Thus, when the infinite is to be summoned in response to a problem of decision, a device that affords the breaking of this symmetry must be implemented. This is why traditional cultures appeal to random processes: if the divine is the infinite, then one can access the divine only through random processes, for each *actual* event that explicitly emerges from a random process directly breaks from the symmetry constituted in the virtual equality of all the *possible* events at infinity. Thus, by appealing to a stochastic process, one circumvents the intentional or subjective decision and appeals directly to the *disinterested* objectivity of pure symmetry.[70] The successes and failures of these practices may lay in each one's capacity to reticulate properly—that is, to transduce the preindividualities of the settled actual world.

We are bathing in randomness, and our knowledge is constituted by its contraction and synthesis; our entire organism is a bootstrapping of randomness into self-production. Randomness can be construed as resistance against compression into pattern or recalcitrance against reduction by redundancies or invariances: the random is, in this sense, that which perpetually escapes capture by the reductive model. This can be regarded as an effect of self-organization: the virtual teleology inherent to each self-organizing system (they must constantly reproduce their boundaries) inevitably leads them to distinguish between signal and noise, between relevant and irrelevant, between sense and nonsense. Randomness is that which, by definition, escapes the system's criteria for relevance, because these criteria are nothing more than a model of the system's environment, its *memory*, inscribed in its degrees of variational freedom or the possible system states afforded by the system's intrinsic topology. Thus, when something is perceived as random, it means that this thing is of a complexity unaccountable within the model of reference, something that escapes the system's capacity to take on varied states. And thus we can do no more than sample randomness in its singular expression; we cannot reduce it or contain it. It is absolutely independent from our modes of capture.

We have known since Poincaré and the three-body problem that deterministic systems *can* and *do* lead to unpredictable nonlinear behavior. Deterministic chaotic behavior arises over time from the amplification of the imprecision of our approximations, from the intervals upon which our measurement segments necessarily fall. In the continuous spaces of classical dynamics, measurements necessarily fall upon intervals, unavoidably glossing over infinitesimal details that inevitably express themselves with time. Deterministic chaos expresses the unknown differences hidden in the gap, which diffuse through state space, resulting in the topological mixing of trajectories (the chaotic trajectory, in time, gets arbitrarily close to every point in the system). In the paradigm of quantum mechanics, a second model for randomness is observed in nature: a pure indeterminacy. In quantum systems, the randomness is thought to be "intrinsic": there will never be a way to predict the events themselves, as Einstein would have wished, but only the probability of there being an occurrence of such and such a type at this or that location in spacetime. Furthermore, in biology, things get magnitudes messier. These two types of randomness participate and propagate across organizational strata in top-down, bottom-up, and transversal proliferations that translate boundary conditions throughout the nested levels of organism in what Longo and Marcello Buiatti have called "bio-resonance."[71]

In any case, we know that the entry of randomness into the system is not limited to the intervals of measurement and the limits of our bounded rationality. Just as in the Bamana divination's use of deterministic scrambling techniques to obtain randomness, computer science has discovered that unpredictability can emerge from the simplest rule-based algorithms. This is known as *pseudo-randomness*. Though defined by simple algorithmic rules of sequence, programs like Stephen Wolfram's Rule 110 cellular automaton can nevertheless take off in their own private dimensions, exploiting the gaps afforded by the digital system to found an unpredictable and perhaps *creative* behavior. Wolfram speculates that such algorithmic systems that lead to undecidability and support universality are in fact more common in the universe than those that are decidable. Undecidability is a factor of what he calls "computational equivalence." Along the same lines as what we have been suggesting regarding the relativity of randomness to a given model of objectivity or frame of reference, he reasons that there is an upper *limit* to the complexity a computation can have, for as soon as a program is "Turing-universal," meaning as soon as it can simulate *any* other program, just like a Turing machine can, then, being *universal*, there is nowhere *more* complex for it to go: it can now potentially simulate entire universes if given enough time and resources. Hence, all Turing-complete systems are computationally equivalent. If these algorithms, these sequences of operations under given constraints, produce unpredictable, nonterminating behaviors, Wolfram speculates, it is because these computations are of the same sophistication as the ones simulating our own conscious processes. Accordingly, he suggests that objects that appear to human consciousness as intelligible, as objects of understanding, have naturally tended to be computations that were less sophisticated than that of its own conscious process.[72]

Now, it is important to stress the *operational* nature of this account, a feature it shares with Leibniz's linking of *qualia* with the analytic methods for approximating continuous functions. Our organism must meander through innumerable nested levels of environmental complexity to arrive at a given concept or percept. Only trivial instances of thought and reflection ever halt on a distinct solution, only "obviously simple" configurations that are reducible and compressible to more economic patterns. But these instances are rare, minor, and perhaps insignificant exceptions floating in an ocean of indistinction. If we are truly capable of computational universality, then, given infinite time, our minds can in principle compute anything. But most things in nature nevertheless cannot be tractably reduced to the fullest perfection and purest distinction of their initial conditions, because these things too are universal and can compute anything in principle. Rephrased in these ways, Wolfram's conjecture begins to deeply resemble Leibniz's claim. What if sensations, passions, affects, intuitions, *qualia,* and other confused cogitations do not merely result from the failures of distinct cognition, as does obscure knowledge, but in fact reflect real operationally exposed features of our world? What if indistinct cognition results from *real operations* that do not result in a final terminus? Daniel Dennett argued that the patterns we successfully compress out of more complex data sets "are real because they are (somehow) *good* abstract objects."[73] But I would argue that they are no more real than the data sets we cannot compress. If patterns are "good," for Dennett, it is merely that they are *useful* for the cognizer, allowing for an economical use of its degrees of freedom in approximately reconstructing a feature of the environment. But they are no more real than the "bad" random structures that fail to be tamed by such operations of compression, and thus do not serve the cognizer's ongoing domestication of its world. Indeed, in their recalcitrance to our attempts at reducing them to patterns, they might even be said to be *more real,* more *subject independent,* than the features we can reduce. Might we not have in this operational account a way of beginning to naturalize the traditional relation between the lower

all experience results from activities of construction

and higher cognitive faculties, as well as the relation between arts and utilitarian technologies?

If "what I cannot create, I do not know," it is because all experience results from activities of construction. To reiterate, cognition never happens spontaneously: it is founded in action, gesture, posture, and sequences of operations that modify our perspective on the world and expose the constraints of our interface with the real. Experiencing something is adopting the series of postures required for constructing that experience. This understanding of experience as constructed through series of materially instantiated operations is inseparable from constructive interpretations of mathematics. Since Poincaré and Brouwer, these non-Platonistic perspectives on the *construction* of mathematical objects have converged with descendants of the *type theory* Russell and Alfred North Whitehead developed as a means to rebuild logic and mathematics on foundations that avoided the paradoxes of set theory. This convergence is now expressed in the *Curry-Howard isomorphism,* which demonstrates a direct correspondence between computer programs and mathematical proofs. In other words, it implies that the theory of computation can be interpreted as *type theory,* and vice versa. A program is essentially the proof of a certain type: if a certain type has a *token,* then that token can be regarded as being the proof of the existence of that type. Thus, by constructing a given proof, by producing an algorithm that terminates on a given *final* construction, one is instantiating an exemplary element of the corresponding type. In the last ten years, these ideas have seen tremendous success. Building on a history intersecting the BHK interpretation of intuitionistic logic, Church's Typed Lambda Calculus, Stephen Kleene's notion of realizability, Per Martin-Löf's constructive type theory, and proof assistants like Coq and Agda, the recent development of Homotopy-Type Theory, a new branch of mathematics discovered by Vladimir Voevodsky and being advanced as a theory of foundations, now provides potential links between computation and the geometries, symmetries, and geodesics central to understanding the physical world.

If all experiences of the world are constructed through series of operations, it follows that all experience can be interpreted analogously to computability theory. There are series of operations that halt, and there are series of operations that do not halt. There are programs that find a terminus in a distinct pattern, and others that never arrive at such a terminus, and thus remain indistinct. In type theory, these nonterminating programs belong to what is known as the *bottom type* or, somewhat erroneously, the *empty type*—erroneously because it is far from empty: it comprises *all* nonterminating programs and is empty only in the sense that those programs produce no *proofs,* no exemplary elements or tokens.

According to Wolfram, the randomness we observe on the edges of our understanding are the products of computations *equivalent to those that give rise to our own conscious experience.* They seem random only because they are computationally *equivalent* to us. Given the appropriate rules, if even the simplest discrete rule-based systems can support Turing universality, then this universality must be somewhat ubiquitous in nature, which, if true, would vindicate Leibniz's conjecture that indistinct perceptions owe to the real complexities of subject-independent features of the world. We can produce them ourselves (the operating systems in our phones are Turing complete), but when we encounter them in nature, we cannot faithfully reduce them to their initial programs and constraints due to *real,* mind-independent limitations, not just subject-relative constraints, such as those that give rise to deterministic chaos. This is a big deal, because, in a sense, all science is about deciphering how nature progressively scrambles its origins. This challenge is familiar to contemporary physics. The way the quantum level of reality "trickles up" to the classical world we naturally observe is analogous to an advanced cryptographic procedure, many of which are based on large polynomial factorizations. But that universal behavior can emerge from simple rule-based systems is significant because it implies that, even without the measurement intervals responsible for deterministic chaos in classical *continuous* systems—that is, even in *discrete* frameworks composed merely of yes/no options—certain initial conditions will give rise to processes that completely scramble their origins. The ubiquity of universality may ultimately imply that, no matter how well we eventually come to know our world, there will always be doubt as to whether the world we observe is the actual "hardware" of the real, or some "software" running on the "hardware" of the real, or just "software" all the way down, a cascade of nested worlds virtualizing their subworlds ad infinitum. This echoes Luciano Floridi's compelling argument that the question of whether the real is ultimately continuous or discrete is logically *undecidable.*[74]

But even if this speculative theory of universal computing evokes a certain vertigo, Rule 110 is just a very sophisticated pseudorandom number generator, similar to the one implemented by Bamana diviners. In the Bamana divination, a "seed" value given by the four initial symbols will generate 65,535 different verses before it repeats. But we have come to understand that certain binary procedures will *never* repeat. Though Wolfram somewhat deviously tries to claim this discovery for himself, the notion has its origins in Marsten Morse's 1921 reformulation of Axel Thue's 1906 aperiodic binary sequence construction. Long before this, it was known that the algorithms for approximating irrational numbers like the √2 also exhibited aperiodic behavior. And in the deep histories of African divination, if any tradition ever implemented the binary placeholders of a counting system within the practice of fortune telling, chances are they also encountered infinite aperiodic progressions.

Returning to traditional aesthetic concerns, we might be tempted to speculate that the notion of computational equivalence echoes the distinction between "beautiful" and "sublime." If the beautiful corresponds to the agreeable, it can be correlated to those computations Wolfram sees as being of *trivial* complexity, those that result in the objects we take for granted. Trivial programs are those that consistently reproduce predicted results or repetitions and nested fractal *patterns* that we can recognize and reduce to formalisms; they are beautiful in the Kantian sense. Reproducibility, repeatability, and pattern all imply agreeability, a state where cognition "dominates" its perception. And the percept is to sensation what the concept is to thought. Science is the business of describing such symmetries, patterns, and orderings. The computable is equivalent to that which can be *understood,* to operations that *terminate.* And sensations on their own level relax onto their own attractors, the percepts. The "sublime," however, in Kant as in Burke, is that which, by shear magnitude, surpasses our capacities of capture and perhaps threatens our integrity: it does not confirm the existence of our boundary with the greater world. If we think magnitude in the sense of tractability or computability, we find that computational equivalence escapes our understanding by shear aesthetic domination of our cognitive model of the world. This "computational sublime" surfaces again in the number Gregory Chaitin calls "Omega": the probability that a program of N bits will halt. This number, it turns out, happens to be aperiodic, random, and uncompressible. Each bit in its infinite series divides the world into computable and incomputable. And, since each bit in the series is an independent and contingent fact about the world, Chaitin takes it to mean that there ultimately is no intrinsic difference between contingent and necessary truths, for the apparently necessary truths of logic (the computable) themselves seem purely arbitrary (randomness). Programs of *universal complexity* meet the limit of complexity in their capacity to virtualize any other program as a subset of their process. The infinite programs that populate the bottom type transgress the limit of computability. They are thus sublime in this sense: they meet or surpass our computational capacities to reduce complexities to invariances and orderings; like the disinterested cruelty of nature, undecidability and unpredictability are threatening to us because they announce the limit of our capacity to territorialize, to intrumentalize, and to implicitate. And yet, we can contemplate this undecidability only from the *safe distance* of our model of objectivity, founded on comforting symmetries and orderings, just as the character in a Caspar David Friedrich shudders before the expanse of nature from a secure vantage point.

But we must not take this analogy too far. For *both* beauty and sublimity fall under the rubric of indistinct cognition. We are on a different terrain here, and the traditional categories do not map smoothly onto the operational, computational context: we cannot simply align sublimity with universal complexity and beauty with trivial simplicity. For, even the beautiful cannot be truly "dominated" by the cognizer in this sense: as Leibniz suggested, we lack access to a finite account of the operations required for distinctly defining any indistinct percept. A more promising way of interpreting aesthetics in the context of computation is in its potential for extending the Leibnizian intuition of a link between the ineffability of *qualia* and nonterminating series of operations.

Each time an organism probes nature and solicits a response, the organism and the world are essentially enacting a kind of *program,* a series of operations for "constructing" the given experience. Human or nonhuman, an organism can therefore be construed as a type: with its specific cognitive, physiological, and technical affordances, an organism of a given type has access to a specific spectrum of possible disclosures of its world and can construct a given array of possible experiences. The problem is, as Baruch Spinoza famously warned: *we don't know what a body can do.* Experiences are accessible only through doings: perceptions *are* the series of physiological postures that correspond to their construction. And some of those series, we find, do not terminate on definitive proofs: they are processes of construction that do not halt on a final result. The difference between distinct and indistinct perceptions, sentience and sapience, can therefore be understood as reflecting various aspects of the transcendental complexes that locally condition a given agential subjectivity, a given self-organizing process. The operational account of experience exposes that what we have traditionally considered "aesthetic" is in no way inferior to the distinct concepts of "higher" cognition. It is simply that the series of operations that experiencers perform when experiencing indistinction do not halt, do not return a pattern, an answer that confirms the existence of our organism's life-defining boundary.

When the diviner appeals to the unpredictability of his pseudorandom system, or scrambles his initial gestures in pursuit of a symmetry-breaking event, he or she is trying to operationally summon the powers of the undecidable. Artists, for their part, have explicitly integrated these same powers with the objective of circumventing the artist's own determined will, *Pyrrhonistically* withholding judgment by appealing to the forces of cosmic randomness. As a form of technics that privileges aesthesis, and since aesthesis corresponds to the concrescence of differences, art tends to be allied to series of operations that do not terminate, and is poised toward randomness, chaos, undecidability, and indistinction. It requires attitudes and dispositions that are receptive to the unforeseen, hospitable to the contingent and to world-model collapse. Even in works of obvious subjectivity, overt style, or political overdetermination, what ultimately withstands the tests of time is always *that in the work* that neutralizes determinations through operations that avoid halting. For, it is this very feature that affords the artwork what Walter Benjamin called its *aura*, its originality and singularity, its divine ambivalence and ambiguity. Artworks program the human organism to adopt nonterminating series of gestures and postures, and thus generate indistinct cognitions. But their ambiguity, thus described, has less to do with indeterminacy per se than with the nature of the interfering operations that compose the *je ne sais quoi*, their particular way of neutralizing each other's effects such that they keep aloft of the terminus, and *stand for themselves*.

2 DETERMINISM AND DRIFT

Aesthetics and Technical Determinism

Damien Hirst's *For the Love of God* is a diamond-studded solid-platinum human skull. It would be difficult to find a more poignant exemplar of the ostentation that is so often thought to characterize the artistic impulse than this shiny, gaudy jewel fashioned from the rarest of materials into a symbol of impermanence, a *memento mori*. Even the title of the work seems to foreshadow its artistic reception by a vast public who could never hope to own such a luxurious object: "oh for crying out loud, a diamond-studded platinum skull." Another specimen might be Maurizio Cattelan's *America*: "oh for God's sake, a fully functional solid gold toilet." One often hears this kind of chatter in the crowd, this kind of dismissal of the arts as frivolous and wasteful activities. Indeed, it is common for people to think of art in terms of ostentation, and not just when considering works made of solid platinum or gold. In comparison to technical objects, artistic objects are justifiably regarded as devoid of practical utility; if they do have any hint of purpose, it is thought, it could only be to resolve the subjective ambition of the artist or the patron who, through various forms of pretentiousness and obtrusiveness, wants to signal to us their success, their wealth, their ruse, or their *fitness*, whether truthfully or not.

This popular understanding of the social role of artworks intersects the genealogies we discussed in chapter 1. We saw that, at least in humans, sensation and knowledge are entangled and that there is no clear and formal way of distinguishing "pure" *aestheta* and "pure" *noeta*. We encountered several blocks on this traditional assumption, including the historical problem of the analytic–synthetic distinction and the related problem of how to *found* beliefs and knowledge in the "givenness" of experience, and realized that our grasp of the difference between necessary and contingent truths is remarkably shaky. In light of all this, I suggested that we were unwarranted in demoting aesthesis as an "inferior" cognitive faculty, for in fact, there is no way to clearly orient this supposed hierarchy of inferior and superior cognitions. But any inquiry into the conditions of *perceptronium* must investigate not only the logical inconsistencies that haunt our common designations of the purview of knowledge and sensation but also the inconsistencies in our consideration of practical ventures, of technical or applied knowledge. We will thus gain a lot from comparing art with other forms of technology and from considering how our biases condition our interpretation of similar distinctions, such as: useful versus useless, practical versus ostentatious, and central or important versus eccentric, marginal, or irrelevant.

As it turns out, this pervasive tendency in the West for preferring the "superior" cognitive faculty over the "inferior," the distinct over the indistinct, the intelligible over the unintelligible, and sapience over sentience also plays out in a much more commonly held assumption: the complex propensity, in the realm of human practices, to privilege the technical or utilitarian over the aesthetic. It is this inclination, I believe, that induces the common *art-as-ostentation* perspective. So, it is worthwhile extending our investigation into the practical realm to see how the same biases echo through our understanding of art and technics. Indeed, as we will see, it is also reflected in contemporary evolutionary or adaptationist readings of aesthetic preferences and artistic impulses. In this chapter, we will try to undo this prejudice in both the technical and the evolutionary domains. It will become clear that our bias for privileging distinct cognition over indistinct cognition spills over into the practical domain and skews our technical and pragmatic priorities, distorting even our interpretation of the specificity of hominization with regard to nonhuman evolutionary genealogies.

Since the Greeks, the arts have often been viewed as an inferior kind of practice when compared with other human endeavors. Mirroring the claimed inferiority of indistinction with regard to distinction, the arts have been considered useless, ostentatious, and frivolous. Even in the evolutionary narratives that now dominate the discipline of contemporary aesthetics, this same prejudice is expressed in the way aesthetic tastes in humans are suggested to derive from sexual selection in nature, which is typically considered evolutionarily less significant than environmental selection. Environmental selection is thought to be the true engine behind evolution, selecting for survival-oriented characteristics, whereas sexual selection produces only colorful plumage and such other phenotypic expressions related to reproductive fitness. Across the board, aesthetic pursuits are deemed inconsequential, causally inefficacious, and trivial, whether in humans or in our prehuman evolutionary background, and hence aesthetic preferences are considered *by-products* of real adaptive pressures. But, just as we did in the first chapter, it is necessary here to analyze and reassess these assumptions in the practical realm. We will discover that they too rest on beliefs that, upon closer inspection, reveal themselves to be critically flawed, their distinctions fundamentally vague, their premises ungrounded, and their terms inseparably entangled.

From the ancient Greek dismissal of visual arts as mimetic and determined and the correlative appreciation for the leisurely and freely creative power of the poet to the current adaptationist interpretation of art as a product of sexual selection and peak shifted stimuli, or again as an example of "costly signals" in signaling theory,[1] there has been a current in aesthetic thought that has discriminated between art and nonart on the basis of a general economics of energy efficiency. Thus, art has continually been bound up with the concepts of leisure, free will, and ostentation. The distinction between the leisure time of free men and the survival-bound lives of workers frames the earliest theories of aesthetics. A well-known art-historical truism is that, for the Greeks, visual art was considered a lowly activity because it demanded physical, manual labor. Anything fabricated with hands was deemed unfree and determined. And so, the visual arts were from early on excluded from the category of *poiēsis*, truly creative production, and reduced to mere *technē*. By contrast, the poet's inventions were considered products of freedom, as they were unbound to mimetic constraints. Free men, it was thought, didn't produce art, but merely enjoyed it. Suggesting it contributed to maintaining the stabilities of the state, Aristotle argued that art was favorable, even though he considered the artist himself to be a laborer, a trivial part of the social system as a whole. As base activities, he thought, artistic tasks could be carried out by unfree minds and hands. A culture of leisure that included aesthetic artifice was considered crucial to fostering rational thought and publicly engaged citizens. Greek culture segregated the artisan from the man of leisure, believing that the role of the working artisan, much like that of the slave, was to participate in providing free men their leisurely lifestyle: artisans did the work of embellishing the life of the free man, entertaining him, providing him with tragic *catharsis,* to liberate the logos from excesses of emotion, to make it *available* and attentive to public affairs.

With Kant, this same prejudice hinges on an economy of interest and disinterest. Intentions driven by necessity are in essence deprived of the faculty of judgment: one who is starving, he argues, will eat anything. Only those who are assured to have their appetites satisfied can hope to see things with a liberal judgment of taste. According to Kant, aesthetic appreciation, and with it the universality and reconciliation it signals, is possible only once liberation from such necessity has been achieved. The story is the same in anthropological and sociological debates, where, following a historical thread through Thorstein Veblen and Pierre Bourdieu, the tendency has been to categorize types of art and aesthetic tastes according to class circumstance.[2]

It seems fairly obvious that, when one has the privilege of leisure and freedom, one can devote more time and energy to the activities afforded by such privilege. Yet it nevertheless seems wrong to assume that free time and resources are a *cause* for artistic activity; at most, they only account for a condition of possibility within a given circumstantial framework, an arbitrary criterion that, if fulfilled, affords the "extravagance" of a given practice (or nonpractice). They account only for what Aristotle would have called the "material cause" of aesthetic practices. In fact, we can argue that art is more than an extravagance or a privileged affair that can happen only once one's appetites are sated, or once one gains freedom from labor. For in another sense, it *is* the labor; it is the discipline of not submitting to unfounded judgments, of not letting the utilitarian impulse impose itself on reason. Art is a "rigorous laziness": it is a refusal to act on unfounded motives, a rational resistance to the self-preservative function that echoes throughout other forms of *technē.*

Moreover, the historically pervasive interpretation of art as puerile or ostentatious follows from the tendency to establish prejudiced demarcations between serious activities (*negotium*) and leisure activities (*otium*), or to oppose the productive to the unproductive, and such interpretation has been promulgated throughout history's characterizations of art. Étienne Souriau dismantled this prejudice in Émile Durkheim: "It is totally impossible to find in the facts such a demarcation between serious activities, constituting industry, and the non-serious, forming art."[3] In his critique of the hidden prejudices in Sartre and Bourdieu, Jacques Rancière argues similarly.[4] Even beyond the historical avant-garde's critique of artistic autonomy and disinterested aesthetics, or beyond the rational exclusion of the artist from Plato's conception of a republic, curious reversals and displacements of the same implicit *doxa* that apportions tastes and judgments according to their *habitus* are continually exposed.

The lesson to be gleaned is perhaps the following: we should be careful not to confuse our judgments of the circumstances affording a certain phenomenon with the sufficient reason or *cause* of this phenomenon. Adaptationists are mistaken to presuppose a causal link between sexual selection and art on the grounds that both seem to favor aesthetic "extravagances." And though it is true that patronage has always favored the tastes of the patron, and that it is easy to characterize the contemporary art-fair market as driven by ostentation, it is nevertheless important to remember that examples also abound of art that is difficult to reduce to such explanations. *Art brut,* "outsider" art, folk art, mediumistic art, and "primitive" or "tribal" art are domains that proliferate with counterexamples: works produced according to very different motivations and within radically different economic circumstances, works that are born of productive engagements with therapeutic, communicational, ritualistic, or divinatory concerns, rather than mere ostentatious attention seeking or status garnering.

It is tremendously significant, for instance, that "outsider" works are devoid of appeals to worldly themes or historical relevance, that the artist will sometimes produce works in complete isolation, the aesthetic *horror vacui* of the artwork acting as a ballast against the social vacuum in which it is conceived. It is rather the "art world" itself as a trans-institutional cybernetic system of selection and ordering that captures and offers these "outsider" works to history, if at all. Take, for instance, the fifty-fifth Venice Biennale. It was named after the project of self-taught artist Marino Auriti: *Il Palazzo Enciclopedico.* An outsider's dream of uniting human knowledge into a single *encyclopedic palace* became curator Massimiliano Gioni's grand theme for the premier international art expo. The exhibit featured a plethora of outsiders who often did not consider themselves artists; it was rife with folk, craft, and "low-art" aesthetics, in great contrast with the billionaire one percent's yachts docked before the Giardini. This contradiction was brilliantly emphasized by Jeremy Deller's British Pavillion mural of a giant William Morris heaving the Russian oligarch Roman Abramovich's superyacht into the sea. Ostentation can therefore be said to work both ways. If the Biennale takes such a curatorial step, it hints toward a common inversion of the principle of ostentation. Much like the potlatch reverses the principle of wealth accumulation while preserving the logic of economic competition, the showcasing of outsiders in such a lofty context harkens to the bourgeoisie's ostentatious appreciation for the "vulgar" or "naïve" aesthetics of the poor or of the common. In this way, it extrinsically forces historical relevance onto a kind of art that is often *purposively* and *intrinsically irrelevant* in its inception.

We can see the same tendency to arbitrarily distinguish a norm from an exception wherever it is convenient in the production of the "just-so" stories that often characterize academic aesthetics. There is a trend that dismisses what are regarded as the "extremes" of the spectrum, considering them irrelevant. Denis Dutton exemplifies this:

> What philosophy of art needs is an approach that begins by treating art as a field of activities, objects, and experience that appears naturally in human life. We must *demarcate an uncontroversial center* that gives more curious cases whatever interest they have.[5]

But such arguments are misguided. The uncontroversial examples of art do not give the "more curious cases whatever interest they have." Quite the contrary is the case. According to our elaboration of aesthesis in chapter 1, creativity flows in precisely the opposite direction: from those imperceptible, free-floating differences to the recognizable and fixed differences we observe as the "uncontroversial center" of human aesthetic practice. William James was right to note: "Any one will renovate his science who will steadily look after the irregular phenomena. And when the science is renewed, its new formulas often have more of the voice of the exceptions in them than of what were supposed to be the rules."[6] Similarly, Rancière shows that, if the center seems *uncontroversial*, it is merely because it dissimulates the *doxa* of a privileged perspective: the center is always the site of implicit judgments, smuggled in through sleight of hand.

The center itself is perpetually reconfigured by the strange incursions of idiosyncratic novelty emerging from the periphery, from the fringes, from the extremes. In effect, such restrictions of art to an "uncontroversial center" risk limiting our perspective to the "just-so" stories that Stephen Jay Gould warned about. Adaptationist thinking attempts to explain the periphery by virtue of the center, the controversial by virtue of the uncontroversial, when in fact this goes against the central Darwinian doctrine: through natural selection, small differences, extreme cases, and idiosyncrasies lead to the fixed and normalized features we readily observe. As this process happens—and this is precisely what disarms the adaptationist method of reverse engineering, save for coarse-grained just-so guesses about adaptive histories—natural selection inherently blurs the course of its path, much like a cryptographer's cipher scrambles the algorithmic procedure for encrypting the message.

If we want to understand artistic activity and the aesthetic field it interacts with, we have to consider the entire range of activities. It at once becomes clear that the center is of little importance in comparison to the plethora of exceptions to the rule: what we need to describe is the manifold transitions between normative attractors that draw certain aesthetic preferences in and the wild idiosyncratic exceptions that everywhere challenge these attractors. This is what interpretations of art founded on notions of ostentation, disinterest, leisure, or purposelessness collectively overlook. Raw art, outsider art, and divinistic and ritualistic arts are fields of aesthetic activity that should therefore be regarded as privileged counterreferences to the art-as-ostentation perspective. Outsider works, by definition, emerge spontaneously in private contexts, and are thus intrinsically incompatible with the ostentation interpretation, even though they are nevertheless sometimes selectively summoned "upward" by the street-cred-seeking figureheads of "high art" and equivalently by ostentation trying to relieve the guilt of privilege for the millennium's *nouveaux riches* art investors.

The apparent congruence between sexually selected traits and aesthetic artifice is due largely to sexual selection's tendency to become dominant in the absence of environmental selection. More variety is afforded survival in low-environmental-selection contexts. And so, among these various affordances, many specific niches become operative around a variety of aesthetic preferences. We should not be surprised to see art thriving in social contexts of leisure any more than we should be to observe variety thriving in economics of abundance. Aesthetic practices thrive in these contexts simply because they are *afforded* capacities to do so. Freedom is a condition of mobility and expression.

By analogy with the thermodynamic constraints of the physical world, it is rather clear that excess energy allows more variety of organizations. Energy *affords* fluidity: a solid becomes fluid as the heat rises beyond the melting point. The crystal, on the other hand, is a compromise of the substance that has cooled down and for which little energy is available. The molecules in the crystal can be said to exist within an *economy of energetic scarcity*, and thus are afforded little variety: they are constrained to the symmetries allowed by their atomic bonds. When the energy dissipates, crystal molecules are forced to adopt the most optimal organization and crystallize into a unified symmetrical structure.

Analogously, one can imagine the descendants of birds of paradise migrating to harsher environments where resources are scarce and evolving such that they lose their extravagant plumage and are forced to *sat-isfice* more pressing survival-related constraints.[7] It is no surprise that, when environmental forces are such that peacocks will die if they keep their fancy tail, selection produces a less extravagant pheasant. It should also come as no surprise that humans living in extreme survival circumstances might tend to produce less artifice than humans living in leisurely contexts. Like all things, art does have "material causes," and in places where there is no energy and time to spend on such activities, such activities simply do not take place. But this does not mean that art is logically allied to ostentation, leisure, or abundance; our attribution of such distinctions is always prejudiced. Any rigorous analysis should remember that such assumptions risk having us overlook artistic activities in those places where they would seem materially improbable. And if we define aesthetics according to these ostentatious practices, we miss the point.

A very telling example is that of the human adoption of sedentism, one of the most decisive and defining technological transitions in human culture, in effect constituting the bridge between prehistory and history. The belief that sedentism was a direct result of the invention of agriculture has been revealing itself to be a fallacy; quite to the contrary, a grand aesthetic project may have been the driving force behind its development. It has historically been assumed that "civilization" starts with the advent of agriculture and sedentary life. But since the discovery of the Neolithic archaeological site at Göbekli Tepe, southeast Turkey, in 1994, a debate on this question has been rekindled. The site is an example of something no one thought existed prior to its discovery: Neolithic monumental architecture. It is composed of massive stone pillars and carved relief sculptures. Most astonishingly, according to the late Klaus Schmidt, the archaeologist who discovered and excavated the site, it was built by hunter-gatherers four thousand years before sedentary civilization spread through the Fertile Crescent. Since there is little evidence that anyone resided on the site, and due to the abundant traces of hunted wild animal remains nearby, Schmidt suggests that the people who began to build these structures were still hunter-gatherers and not yet cultivating the land and domesticating animals.[8] This is immediately controversial for archaeologists and paleontologists faced with the problem of how a hunter-gatherer society could have built such monuments, presumably without the social hierarchies, economic infrastructures, and technical capabilities required for endeavors on this scale. The discovery echoed a handful of previously published paleontological theories that predicted such a find. The most controversial of these was that of Jacques Cauvin, who argued that a cultural revolution from animism to the worshiping of divinities must predate agriculture and the "Neolithic package." Cauvin suggested that the transition to worshiping divinities could itself have triggered the technical revolution of sedentism.[9]

According to Schmidt, a remarkable revolution in our paleontological account takes place with Göbekli Tepe. Not only does it not fit what we formerly took to be the necessary affordances for monument building on this scale—namely, sedentary living—but the site at Göbekli Tepe may in fact have played an important role in the transition to permanent settlements. Archaeological evidence suggests that the strain of wheat eaten around the world today was first domesticated near Göbekli Tepe, and so some experts speculate that these people may have been the first farmers. This would mark a reversal of the explanations typical of technical determinism: here the technology of agriculture seems to have been adopted *in order* to construct this massive architecture and its many exquisite figurative reliefs.

Thus, this spiritually and aesthetically motivated collective "art project" may have been driving one of the most significant transformations in human history: the passage from hunting-gathering to agriculture and sedentism. Because the construction of the site was so time-consuming, the culture may have had to find ways of making their organization more efficient *in order* to free time and energy to devote to the project, and in this way, progressively adopted the features of sedentary life as an effective solution. The technical challenges this culture had to overcome in order to build these architectures were without precedent and revealed much about abstract thinking, geometry, cooperation, and organization. New concepts had to be conceived; new technologies had to be invented; calculations and measurements had to be resolved. The project may have been so obsessively invested that it required the development of farming *in order* to feed the multitude of workers. Hence, sedentary life, and indeed civilization itself, may be the posterity of an inspired art project. The technology of agriculture may have followed from the aesthetic practice, rather than the reverse. Sedentary life may have proceeded from monumental architecture, rather than the reverse. Civilization may be a by-product of art. With Göbekli Tepe, a spiritual and aesthetic practice may have taken precedence over survival-oriented tasks in a context that apparently did not favor this extravagant endeavor. Given our usual prejudices about the nature of aesthetic practices and their material causes, we would not have assumed this to be possible.

It is true, as the anthropologist E. B. Banning warned, that we must be careful not to be misled by our own cultural biases, which sometimes favor falsely reading strict distinctions between sacred and profane spaces into early cultures. He reasons that it is moreover highly improbable that the culture at Göbekli Tepe was not already domesticating at least some animals and plants.[10] But it is nevertheless clear that the site represents several layers in the Neolithic *transition* from nomadic to sedentary life, and perhaps from animistic culture to divinistic culture. The ambiguity and controversy around the site and the questions it raises about our modern cultural biases are inseparable from our prejudice for reading cultural revolutions as being necessarily driven by technological affordances. They point to a need for a better understanding of how aesthetic and technical aspects of human culture are functionally interrelated. While we argue about which came first, cultural revolution or technological revolution, we are overlooking the more essential character of the problem. Technology and culture advance together in an oscillating metastable fashion, each responding to the other, neither one definitively determining or causing the other. Both sides of the coin are, in the final analysis, transductively constructed and reciprocally enacted. The culture that built the monument did live in what was a quite fertile environment rife with resources and food, but nevertheless, these people lived at a time when some of our ancestors were still abandoning caves.

So it seems we were misled in our assumption that only sedentary culture afforded the capacity to build monuments of this type. We generally tend to think that socioeconomic forces and technical innovation drive human progress. But, in this case, evidence may be suggesting that a culture was driven to build monuments *before* they had achieved what we (until now) took to be the prerequisite conditions: a surplus of population density and people-power that resulted from sedentary life and agriculture. This is reminiscent of the notion of the "black swan": a black swan is always thought not to exist (or to have very little probability of existing), until it is found and subsequently absorbed into a new anticipatory horizon. In the same way, we tend to have a bias for assuming that, in the absence of that which we take to be the material or technical affordances of a particular occurrence, the occurrence is highly unlikely, or even impossible. In other words, we tend to reify the material cause within our system of reference (our model of objectivity), and thus we often tautologically *find what we expect to find*. Clearly then, this bias has an obstructive effect on our evaluation of aesthesis.

The evolution of the arts, of *aesthetic schemata,* is thus often erroneously described as taking a back seat to advances in science and technology. But if Plato and Aristotle were concerned about how the arts promoted or discouraged rational thought, it is because aesthetics and artistic inclinations were known to have effects outside of the strictly sensorial domain. If Leibniz's science and philosophy was, according to Gilles Deleuze, influenced by the baroque aesthetics of his day, it is because the pursuit of knowledge is infected by the greater aesthetic environment in which researchers do their work. No one lives in a vacuum. Great scientists, notably Einstein and Paul Dirac, reported pursuing theoretical formulations for their formal "elegance" or "beauty." Moreover, in the domain of art history, we know that the knowledge required for the invention of photography (photosensitive chemicals, camera obscura, optical lenses) was available centuries before photography was invented, and likewise that techniques for reproducing linear perspective in two dimensional images were known by the Greeks long before their reintroduction into Renaissance Italy by Filippo Brunelleschi. In these examples, it seems that mere technical knowledge was not itself decisive in the coming into being of the new technology; rather, it looks as though these technologies were introduced in response to aesthetic desires, or perhaps more accurately, to hybrid *epistemaesthetic* desires.

Strictly technical affordances, in other words, tell us little about how significant historical events in aesthetics and art come about. Of course, aesthetic activities just don't happen where they are not "afforded" by the epistemic and technical degrees of freedom of a given culture. But it is telling that humans have even been known to compromise their means of survival in the course of symbolic, spiritual, and aesthetic pursuits. In the most extreme cases, they have taken the aesthetic production of spiritual artifice so seriously that their practices seem to contradict their most basic survival instincts. The native peoples of the northwestern coasts of America once participated in great festivals of ostentatious giving, where partakers demonstrated their capacity to rid themselves of possessions. Participants gave away their material belongings, and in some cases destroyed their villages or slaughtered their slaves, in public exaltations driven by the competitive display of divine *detachment* from earthly things. According to a commonly held interpretation, the people of Easter Island may have devastated their island's forests in order to erect their stone monuments, effectively killing the ecosystem, and hence their own means of survival. Are these not examples of purposeful purposelessness? These practices act against the "selfishness" of our genes, which only want us to survive and procreate. To be sure, the Haida and the builders of Göbekli Tepe lived in contexts of relative abundance and material surplus, but this does not imply that aesthetic- and nonsurvival-related activities are wholly technologically determined in the final instance.

What is true, perhaps, in evolutionary terms, is that only those extravagant, ostentatious, or excessive investments into artificiality *that are simultaneously also accompanied by the favorable environmental, technical, and economic conditions* will be forwarded by evolution or history. This is perhaps why there are few Göbekli Tepes in the world and many more examples of *post*-sedentary monument-building cultures, and why the "uncontroversial center" of artistic output is occupied by "aesthetically pleasing," fragile, rare, or ostentatious specimens: sedentism more readily affords monument-building capacities; the rich patron supports the artist's activities. However, the point is that we should not take material causes and technical affordances as ontological attributes of these activities, for they may in fact be motivated and determined by "black swans," contingent and unpredictable occurrences not so easily defined in terms of given uses or reasons. In *On the Gradual Production of Thoughts Whilst Speaking,* Heinrich von Kleist speculated that the French Revolution had been triggered by some irrelevant contingent occurrence: "It was perhaps the twitching of an upper lip or an equivocal tugging at the cuffs that brought about the overthrow of the order of things in France."[11] This example again shows that we can never know which affordances are present until they express themselves in the appropriate emergent circumstances. It is always easy *après-coup* to retroactively trace genealogies according to chains of material affordances. The problem is that, in offering these explanations, we are usually prejudiced in favor of some just-so story, whatever it may be in each case.

This discussion intersects a tension that has been at the center of traditional art-historical discourse itself, that between materialist and idealist perspectives on art. We might understand this through the lens of the Semperian–Rieglian debate. The German architect and art critic Gottfried

Semper had proposed a materialist interpretation of the evolution of artistic styles that was founded in the interface between human skills and material affordances. He proposed that all architectural styles were derivations of originary engagements with the essential properties of the materials interacted with by early humans. As skills improved and technologies evolved, different aspects of materials were teased out of them, but they never ceased echoing these original engagements of humans with matter. The evolution of stylistic themes in aesthetic practices was thus the expression of an ongoing interface between human skill, know-how, and, Semper emphasized, the intrinsic properties of materials. All aesthetic motifs were thus derivative of technical engagements with material properties: crosshatch designs on ancient pottery, for instance, were held to be descended from the practice of basket weaving, where the material constraints of the weave themselves produced such a design element as a necessary by-product.

Alois Riegl would later reject this interpretation. Influenced as he was by the idealism of Kant and Hegel, he insisted that artistic ventures could not be reduced to material affordances and technology. He developed the influential concept of *Kunstwollen*, the will to art, and cultivated the idea that artistic creations stemmed from the spirit and were directed toward final causes and purposes, rather than merely determined by technical and material affordances. Through the lens of art history, which importantly, for Riegl, included the minor and decorative arts, one could, according to him, gain insight into the *Kunstwollen* of a given race or culture at a given time and place.

This idea that certain aesthetic styles represented racial differences would later come to influence Nazi narratives on aesthetics through the work of the second Vienna school, notably Wilhelm Worringer and especially Hans Sedlmayr. We could today view the errors and oversights of this spiritualist school of thought on art as part of the general mishaps of modern historicism, the tendency to synthesize historical processes into "grand narratives" of social progress toward European ideals. These mistakes have widely been criticized in postmodern and poststructuralist thought, and indeed, in part due to Sedlmayr's affiliation with the Nazi party, this tradition fell into disrepute with the postwar rise of Marxist, leftist art-historical readings, which leaned toward materialist interpretations of the genesis of styles. I agree with these critiques: we must strive toward a staunchly naturalist reading of aesthetics and the evolution of artistic styles. However, in light of the epistemaesthetic convergences we have been exploring, I do believe that something in this idealist legacy is nevertheless important to keep in perspective.

Indeed, epistemaesthetics compels us to carve out a middle way between materialist and idealist ways of thinking about art and the evolution of aesthetic styles. With the epistemaesthetic realizations we have seen in the science and art of the last century and a half, this traditional distinction has been shown to dissolve. Realizing that there is no firm ground on which to found a distinction between appearances and truth, or between perceiving and knowing, involves recognizing that consciousness, mind, spirit, and will cannot be divorced from the material world. Indeed, they are materially constituted; they are peculiar manifestations of matter, of causal forces interacting in space and time according to deterministic constraints. At the same time, this does not mean that *traditional* materialism wins over completely on the spirit: after the discovery of quantum mechanics, it is clear that matter can no longer be thought of as a static and inert substance with fixed attributes. This traditional view has been overturned by a theory of matter as *fluctuation,* pure unrest in the depths of the void. Furthermore, physics today views matter as inseparable from the concept of *information,* ultimately implying, following Claude Shannon's theory, that matter is somehow derivative of observation. As we will see in chapter 4, this is why observation bias effects are found everywhere in physics, especially in quantum mechanics, where the role of the observer and the experimental apparatus is integral to the theory. As the American physicist John Archibald Wheeler once succinctly put it, the universe is constructed "it from bit," and the scientist is thus a participant in the evolution of the cosmos. This realization in quantum physics was analogous to what had already been accepted within the context of art history. In reference to the motives of postimpressionist art, Ernst Gombrich insisted that "there is no reality without interpretation," and indeed, this is precisely the insight that precipitated the artistic vanguard's abandonment of illusionism in favor of abstract, expressionistic, conceptual, and experimental approaches. Artists abandoned the representation of objectivity because they had discovered that their true role was not to imitate nature, but rather the reverse. If "life imitates art," as Oscar Wilde argued, the artist's *Kunstwollen* acquires an agential role in the constitution of objectivity, just as does the observer in quantum mechanical experiments. The *Kunstwollen,* itself a functional instantiation of matter, is an aspect of the experimental *apparatus* making the measurement, and is thus entangled with and participating in the bringing forth the material's attributes.

Between the Semperian and the Rieglian perspectives, therefore, the epistemaesthetic approach must carve out a naturalistic theory of the evolving transcendental conditions of artistic and technical innovation. This implies recognizing, against Riegl, that creativity is itself material and that matter is creative, while also recognizing, in some sense against Semper, that this involves embracing an *irreducible* form of unrest intrinsic to matter. By extension, we must think of the historical genesis of art and technology, aesthetic schemata, scientific paradigms, and other functional cultural assemblages as continually *coevolving*. Just as we cannot completely dissociate *noeta* from *aestheta* on the mental stratum, the evolution on the social stratum of these aesthetico-techno-scientific schemata must be described in tandem, for there is no doubt that they must overlap, intersect, and influence each other in various complex ways. Accordingly, for Gilbert Simondon, "there is a continuous spectrum that connects aesthetics to technics."[12] There is of course no denying that *technē* is intimately linked to human evolution: each new technical innovation comes with new affordances and constraints that greatly affect the course of hominization. What we often overlook are the subtle contingencies and underground motivations that participate in the development of these technologies at every step and that the true labor of innovation is not the application of utilitarian reason, but rather the application of a Pyrrhonistically pure reason, an operational mobilization of the *discipline* of *epoché*.

Underdetermination and the Adaptationist Scheme

Let us now take a closer look at how these prejudices against aesthesis as the "lower cognitive faculty" play out in the domain of evolutionary aesthetics. Here the development of aesthetic preferences and practices is held to progress according to the logic of Darwinian selection. The field elaborates a naturalist conception of aesthetic change, connecting human arousal and aesthetic taste to ancestral and prehuman selective pressures. It is clear that human aesthetic inclinations are strongly determined by our genetic legacy, reinforced over millennia of evolution, most notably, according to neuro-aesthetics, during the Pleistocene age. Our aesthetic attractors (tastes, preferences, palates, etc.) are considered to be either those our genetic ancestry will have favored and reinforced through various processes of selection or those that will have been forwarded to us as secondary "by-products" of these active selection processes. Aesthetic preference, like human cognition more broadly, is under constraint by our physiology, our genetic coding, our evolutionary background. This will not be put into question. What our investigation will show, rather, is that aesthetic pursuits are less about the distinction between "useless and useful" than they are about that between underdetermination and overdetermination. Whenever we try to specify the determinations of a specific evolutionary expression, we are no doubt biased to read the correlated determinations through the lens of our own biases: we always overdetermine, *après-coup*, that which reveals itself to be essentially underdetermined when considered rigorously and with appropriate nuance.

Let us begin with the so-called "savannah hypothesis." It shows a statistical tendency in human populations to prefer savannah-type landscape representations over other types of landscapes. This statistical preference in the population is linked to the selective pressures faced by our Pleistocene ancestors. Preference for the aesthetics of savannah-like landscapes was progressively reinforced over thousands of years, as those individuals who were "attracted" to deserts and jungles tended to suffer the pressures of these more hostile environments, while those who were "attracted" to the savannah found within it abundant food, access to shelter, few predators, clean water sources, and so on. The latter were hence more disposed to forward their genomes to future generations.[13] Consider also the "good gene" argument, meant to explain such things as the universality of attractiveness rankings for human faces. Research shows that the closer a face is to a prototypical average, the more it will be judged to be attractive because, the explanation goes, it signals reproductive fitness. In other words, the more one looks like everyone else, the more one is beautiful, because beauty is simply a marker of the "uncontroversial center," the mean. This is the gist of the *adaptationist* interpretation of aesthetics: through adaptation, our genes program us to prefer certain aesthetic objects, situations, and compositions while reacting to others with aversion. It is the idea that our aesthetic preferences are ultimately determined by adaptive pressures. The adaptationist will tend to interpret aesthetic tendencies as products, and especially *by*-products, of adaptations to specific environments and niches.

The trend in evolutionary theories of art, for example those by William Hirstein and Vilayanur Ramachandran,[14] Colin Martindale,[15] and Dutton,[16] is to see artistic expression, and its status as an independent field of human activity, as an epiphenomenal effect of *sexual selection,* a by-product of several sexually selected adaptations. This implies a certain set of constraints they claim characterize the significant vectors of aesthetics. Artistic activities and aesthetic tastes are not directly *survival-related,* and are thus understood as by-products of sexual selection. Beauty, they argue, is associated with rarity or delicacy, because aesthetic tastes are handed down from our adaptations and we tend to be aroused or excited by those features of human behavior that *appear to be independent from survival-oriented purposes.* There is an implicit sense in which this view regards the products of sexual selection as naturally "ostentatious," in that they privilege and favor non-survival-oriented (or indirectly survival-related) traits geared toward impressing peers with extravagances. They are vaguely reminiscent of Aristotle's linking of leisure with the consumption of the aesthetic, Kant's stance on the disinterested experience of beauty, and avant-garde critiques of bourgeois aesthetics and artistic autonomy. It is also a historical thread that fuels the common cliché of the *dilettante* artist. Indeed, this line of thought follows the same argumentation that would have dismissed the possibility of Göbekli Tepe.

Charles Darwin famously wondered what survival benefit was afforded to the male peacock by its extravagant plumage. He apparently worried it might point to a flaw in his theory until he realized that it signaled a separate selection principle was at work. As it turns out, there are many nonsurvival-oriented traits that are teased out of evolutionary phyla by the principle of sexual selection. The songs, dances, and plumage of birds of paradise and the architectural propensities of bowerbirds are oft-cited examples of nonhuman *aesthetic artifice* apparently generated and reinforced by sexual selection. These are understood as an effect of the particular environmental predicament of the species: these birds evolved in isolated regions (typically islands) that are almost devoid of mammalian predators. They live in a context of abundant resources and of very minimal environmental selection. The environment, in their case, is not what predominantly selects which individuals will contribute their genes to the next generation. Here sexual selection takes over as a determinant factor over survival selection. Sexual selection often stresses the differences between the sexes wherever either the males or the females must compete with each other to win access to mates. In most species of birds of paradise, the males compete for females with their spirited performances and acts of seduction. The sexes thence evolve to be completely segregated: they have different functions and activities and different physiological traits that go far beyond their initial basic difference in reproductive organs: extravagant plumage, complex mating songs and dances, intricate architectural constructions, and so on.[17] The birds form no couples: once a male has won the seductive competition, he generally wins over all the females, and the other males have to wait for the next mating season for another try.

Then there is the paradise crow. The males of this species have no extravagant plumage, nor do they perform courting songs and dances or build colorful architectures. Instead, they form monogamous couples with females and share with them the burden of rearing offspring. The males and females look the same and behave similarly. The reason for this is perhaps attributable to a genetic legacy that had their ancestors confront stronger environmental factors, and thus more survival-related selections. For example, in contexts of relative nutritional scarcity, the environment exerts more pressure on the chances the species has for survival, selecting only the best foragers and the most cooperative couples. In this context, there is no time or energy to spare for hedonistic artifices: the species must expend its efforts in feeding and supporting offspring. Each individual and (once they have reached the adult age) each couple competes against the forces of environmental selection, rather than against other members of their sex for access to mates.

Hence, from the adaptationist point of view, the economics environmentally afforded to the species heavily affect which selection, sexual or environmental, will predominate. Economics of scarcity generally lead to the predominance of environmental selection and more "efficient" uses of resources geared toward survival. Economics of abundance, on the other hand, lead to the predominance of sexual selection, which in turn leads to extravagant and ostentatious uses of resources in non-survival-specific activities of seduction. Thus, many in the field of evolutionary aesthetics speculate that this feature of the Darwinian paradigm offers a model for understanding the distinction between art and other, more useful human practices. Art and aesthetic, figurative, poetic, or decorative uses of resources are, according to this interpretation, expressions of intrinsic propensities for individuals to impress peers and garner attention from prospective mates, propensities that become dominantly selective in situations that are less determined by environmental factors. This again mirrors the reading of aesthetics according to the distinction of interest from disinterest, necessity from freedom, or purposefulness from purposelessness, echoing a long-held paradigm of art-historical discourse.

We should note here that the theory of sexual selection has not always been taken seriously. Immediately after Darwin's positing of the theory, the cooriginator of the theory of evolution by natural selection, Alfred Russell Wallace, attempted to diminish its import. Though Wallace could not completely dismiss the possibility that sexual selection did in fact take place, he argued that it was rare and probably had only trivial effects on the overall evolutionary process. Reflecting a typically patriarchal bias, he was especially critical of the idea that female mate selection could be responsible for evolutionary change. His argument succeeded in persuading many, and as a result, mainstream evolutionary biology all but dropped the concept of sexual selection for a hundred years, until interest in it was rekindled in the 1970s with Amotz Zahavi's *handicap principle* and the subsequent *signaling theory* of evolution derived from it. However, even this renewed interest in mate selection mechanisms was perfectly congruent with Wallace's argument against Darwin. The only reason there is sexual selection, this theory assumes, is that sexual preferences have been selected for in the first place by environmental pressures, and so these preferences are therefore still derivative of environmental selection. The females are not "steering" the course of evolution through their aesthetic selection of mates, because even their own preferences are ultimately specified by environmental factors. In this way, seemingly wasteful signals such as colorful feathers and elaborate dances are seen as "honest" signals of fitness. Sexual selection is thus held to fold into natural selection: it is just a minor submodule, as it were, of the overall mechanism of adaptation through the constraints of natural selection.

Now, this view is complicated by the fact that, even though adaptationism is still very much in vogue in evolutionary aesthetics, the adaptationist doctrine of reverse engineering has repeatedly been critiqued within the field of theoretical biology over the last sixty or so years. It is worth taking some time to show that, much as we saw in chapter 1 with regard to *aestheta* and *noeta,* the terms on which the adaptationist doctrine rests are *entangled.*

1. The principal challenge to adaptationist reverse-engineering has been well known for decades: *genetic drift* plays a nontrivial role in evolution. Since populations are bounded and finite, the initial genetics of a population (i.e., migrants breaking off from a larger group) may be contained in a relatively small group of individuals. But, as there is always a statistical "sampling error" in the copying of alleles from one generation

to another, a random drift occurs that diversifies and differentiates the genetics of the group as they naturally diffuse through the allelic state space, even in the complete absence of any specific selective process. Furthermore, some of the randomly drifted alleles will become fixed at locus points and may become resistant to selective pressures under specific circumstances (ratio between population size and the intensity of selection).[18]

2. It is now known that development plays an important role in the production of the organisms we observe, which makes it difficult to associate genotypes with phenotypes. Without a full account of how the ongoing development of the organism within its environment affects its morphogenesis, an unshakeable sense of doubt is cast upon all postulations of genotype–phenotype causal relations. Similarly, it is known that gene expressions are *correlated*. Genes are linked together in complex reticulations and interdependencies, which means that it is difficult to say whether an expressed gene is the result of an explicit adaptation or a secondary effect of another adaptation. This is exemplified by Gould and Richard Lewontin's famous spandrel analogy: a spandrel, which is the tapered triangular shape that necessarily appears in the corners of cathedrals wherever a dome is set atop of arches, is a by-product of the geometric constraints between the dome and arches. If we take the spandrel as the motivating principle of the architecture, we misrepresent the architectural theme of the cathedral, taking the by-product for the product: surely the goal was to set a dome on top of arches, not to erect impressive spandrels in stone. But, when we observe a phenotype, how are we to know whether it is the result of a specific selection by an ancestral environment or the by-product of another unknown adaptation?

3. Another thorn in the side of the doctrine of adaptation is that of stasis and punctuated equilibria.[19] This theory partially echoes earlier theories of saltation and "macromutation," for instance Richard Goldschmidt's "hopeful monsters" hypothesis, where evolution is driven by spontaneous large-scale variation in organisms. Henri Bergson, in his critique of Darwinism, was also highly interested in the prospect of such sudden variations or alternating periods of mutability.[20] Though the macroevolutionary theory of saltation has been put to rest rather conclusively, it is nevertheless well accepted that, over geological time, species do not evolve in tiny progressive increments, as Darwin believed. The fossil record shows that most species will change very little even in

the face of important environmental changes, including transitions in predatorial variety and food sources, and that this evolutionary stasis tends to be *punctuated* with periods of rapid change, implying extinctions and speciation, which often seem uncorrelated or out of phase with environmental shifts such as ice ages, planetary disasters, and so on. Because of the looseness of the connection between speciation and observed environmental constraints, it becomes difficult to link a morphological trait or behavior to any specific environment or context.[21]

4. Finally, another important issue is that selection and adaptation do not take place on a single playing field: each basin of selection is nested within other selecting systems. There is thus a whole hierarchy of nested selection principles at work in each given adaptation. Furthermore, Niles Eldredge showed that the regimes of ecology and evolution are "divorced from each other," meaning that there are two distinct classes of hierarchy that condition each given organism in parallel.[22] On the side of the environment and ecology, the genomic organism is subject to conditioning from its avatars (or actual incarnations of the genome within an individual), these avatars are conditioned by their local ecosystems, and these local ecosystems by their wider regional ecosystems. On the side of evolution and genealogy, the organism's genome is contained and conditioned within the individual's own developmental individuality. The individual is conditioned by the *demes* from which its immediate ancestry has emerged, the demes are limited and constrained by the species and its boundaries, and the species is itself constrained by more general phyletic constraints such as *body plans*. This nested aspect of the selection units and the causal relations between the reservoirs of information they select from make the correlation of a specific trait to a specific adaptive pressure that much more difficult. It also leaves open the possibility of *substrate independent* selections, such as symmetry breakings in statistically "null" regions, which result in "emergent" expressions that are irreducible in practice to the specific pressures of the lower selective basin.

It should be stressed that these difficulties are only exacerbated by the adaptationist view's hypostatization of the *dualistic* distinction between organism and environment. For indeed, the selective manifold can be said to extend from the absolute unilateral determination of the organism's aesthetic preferences by environmental pressures to the aesthetic preference's absolute irreducibility to and dislocation from environmental pressures. In these latter statistically null regions, one can no longer strictly distinguish between organism and environment. This echoes the theory of autopoiesis developed by Humberto Maturana and Francisco Varela:[23] a functional loop that dissolves the strict boundary between the organism and its world, where "living beings and their environments stand in relation to each other through mutual specification or codetermination."[24]

In such regions of dislocation between basins of selection, neither the organism nor the environment comes first: both terms of the relation are "enacted." Hence the organism is not *adapted to* the environment in the sense of having to account for its own determinant influences, for that would require the environment to come first. Rather, the forms and expressions recognized by an external observer emerge in the mutual interaction *between* many levels of constraint and influence through the shared history of codetermination or mutual specification: a perpetual *enaction* of the holistic system. Susan Oyama similarly notes that "form emerges in successive interaction. Far from being imposed on matter by some agent, it is a function of the reactivity of matter at many hierarchical levels, and of the responsiveness of those interactions to each other."[25] Such feedback loops occur in those regions where an emergent dislocation between levels of selection occurs, where the pressures of selection encounter a null region, where there is no way of telling whether a trait was "purposefully" selected or purely random. Such regions are where substrate-independent selection may be said to occur following the reciprocal codetermination of the multiple perspectives composing a world. This can be thought of as the evolutionary echo of the principle of *computational equivalence* we encountered in the last chapter, where aesthetics meets incomputability in the *qualia*-infinitesimal isomorphism.

Simondon reasoned analogously that the engine behind the processes of individuation was *transduction*. Defined as a process that produces its own terms, transduction can also be thought of as an enaction of the relation. The terms do not preexist the relation. They are perpetually self-defined in the toils resolving tensions between various preindividual tendencies. Lynn Margulis's groundbreaking corroboration of the *symbiogenesis* of the eukaryotic cell is a testament to such processes of transduction in nature: the cell emerged from the progressive symbiosis of the organelles it is made up of, which individually evolved from different prokaryotic cells. Deleuze and Félix Guattari discuss the interweaved evolutionary destinies of the wasp and the orchid to the same effect: two heterogeneous phyletic genealogies that progressively intertwine into networks of mutual influence and codependence.[26] The wasp and the orchid are in a perpetual state of *mutual presupposition,* as their parallel evolutionary courses *enact* each other's affordances, drawing them forth into actualization. Everywhere in nature, organisms are "structurally coupled" to each other. Ecosystems are horizontally, hierarchically, and transversally organized meshes of these interactions between self-organizing systems. Each milieu is decoding others and being decoded by others, resulting in the characteristic rhythm of the ecosystem.[27] Each organism is embedded in ongoing, perpetually and mutually enacted sets of relations and constraints with other organisms, as well as with the features and affordances of the environment. We are always too tempted to reduce the turbulent magmatic rhythms that condition the provisionally fixed terms of the world to a single arborescent genealogy, when in fact, a *family tree* is merely one of many possible cross sections of the *family rhizome.* This is the radical implication of enaction, transduction, and what Deleuze and Guattari describe as *transcoding* between milieus: there is not simply one history leading to the organism at hand; there are always parallel stories that need to be told. For each provisional present is grounded in infinite different causal lineages ramifying from infinite different pasts.

The links we establish between phenotypes and genotypes or allelic expressions and selective pressures are therefore, at best, what Gould and Lewontin characterized as "Panglossian" (Dr. Pangloss being Candide's "optimistic" mentor in Voltaire's famous novella).[28] Many of the traits we willfully recognize *as* adaptations are actually by-products of effective adaptations; no one disagrees with this point. But what is important to keep in mind is the difficulty and perhaps impossibility of saying with any certainty whether a recognized trait can be associated with its supposed purpose, its supposed "use value" for adaptation, and thus with its selection worthiness within an evolutionary context that remains, for the most part, undefined. Nowhere is there to be found an objective primary selection principle: each adaptation results from a synthesis and integration of many selection principles. Hence, there is no straightforward way to know which trait is a primary result of selection and which is a secondary by-product (or a tertiary one, for that matter). We can assume spandrels are not primary in the architecture of a cathedral, but in the context of nature, we cannot assume to have ever found the real or primary "purpose" of a specific adaptation, meaning the environmental pressure the adaptation responds to or satisfices. Thus, we cannot say that aesthetic preferences and artistic impulses result from a by-product of adaptation (or of a super-normal or peak-shifted stimulus response), since there is in fact no way to say for sure what the primary "purpose" of any trait is, if there be such a purpose in the first place. It is, incidentally, highly dubious that any trait has any such specific purpose: individual traits always satisfice multiple constraints, and these cannot be exhaustively enumerated. Such problems are intractable in practice (if not in principle), again recalling Leibniz's linking of *qualia* with indistinction.

Another angle from which the evolutionary analysis of aesthetics tends to produce Panglossian just-so stories is through the conceptual lens of the *supernormal stimulus*. Supernormal stimuli are a specific kind of evolutionary by-product. Due to evolution, our bodies and brains are wired to induce the required survival-affording responses to environmental stimuli. We are disgusted by the smell of rotten food, aroused by sexual imagery, and find symmetrical faces attractive because the survival of our ancestors depended on eating well, procreating, and choosing fit mates. We must accept this general coarse-grained adaptationist maxim. Our genetic background has hard-wired us, literally, in such a way as to be inclined to do that which favors our survival and procreation, and these hard-wirings also give rise to by-products such as supernormal stimulus responses. First coined by Niko Tinbergen in the 1930s, "supernormal stimulus" refers to environmental triggers that artificially stimulate an organism's receptors *more intensely* than the natural stimulus the receptors evolved to respond to. For example, the satisfaction we reap from eating fat and sugar may have driven our ancestors to find precious sustenance for millions of years; today however, in the food deserts of inner cities, dotted with fast-food restaurants, candy-laden corner shops, and advertising specifically designed to trigger subconscious responses, these same receptors work against healthy human survival. The hypersalty, hyperfat, hypersweet ingredients in processed and fast foods are *supernormal* stimuli for our natural receptors, as are the giant close-up images of mayonnaise dripping from a hamburger on a billboard. These, it is assumed, arise as adaptive by-products that overextend our adaptative inclinations to seek out calorific sustenance. Every time a new genetic recombination gives the organism an environmental advantage, it also opens the organism up to new potential *super* stimuli. We can think of every adaptation as open-ended: each new change comes with hidden affordances that will express themselves only given the right circumstances. It is as though each of the actualized traits that make up the bare machine of the organism extends out into a plethora of virtual connections and potential reconfigurations of the topology of its relation with the environment but these connections and reconfigurations are actualized only when the correlative *Umwelt* is participatorily enacted.

The supernormal stimulus can therefore be taken to imply that human propensities may be, in many cases, driven by these infraceptive evolutionary effects and triggered when specific yet previously unknown circumstances arise. If this is the case, could it be that our ancestors first painted the walls of caves in a *supernormally* stimulated trance induced by artificial images? Indeed, such an assumption is behind Ramachandran's claim to have discovered the "deep structure" of art.[29] According to him, the reason that the prehistoric Venus sculptures, some of the earliest known human art objects, have exaggerated breasts and female curves is that the enlargement of these sexual identifiers acted as a supernormal stimulus for the persons crafting them or admiring them. The prehistoric Venus is but one example. There are published results, for instance, of experiments suggesting that supernormal traits, such as enlarged eyes and lips, are ubiquitous in artistic representations of the human face.[30]

These phenomena are generally explained as progressively *peak-shifted* stimulus responses. The peak shift is an effect observed in animal *discrimination learning*. This is where an animal is *trained* through reinforcement to discriminate between stimuli. Train a rat to prefer slightly rectangular shapes over squares by providing food where the rectangles are but not where the squares are and then show the rat a shape that is more acutely elongated and rectangular than what it has been trained to prefer. The rat will now respond more excitedly and intensely than with the original rectangle. The difference between this and the supernormal stimulus is that it is a *learned* reaction, not an instinct: the peak response to the stimulus has shifted through training. Ramachandran provocatively describes the role of peak shift in art with the phrase "all art is caricature."[31] He speculates that all art, and indeed all aesthetic experience, can be explained by our physiological responses to peak-shifted stimulus values.

Drawing from the concept of *rasa* ("essence") in Indian aesthetics, he suggests that, when artists attempt to tease out the essence of that which is depicted in their work, they will, consciously or unconsciously, either exaggerate the features associated to that essence or, conversely, diminish those features that detract from or confuse the message. Accordingly, he claims that the large breasts and hips and unnaturally narrow waists of the figures featured in Indian Chola sculptures are supernormal stimuli for evoking the female body (one could make a similar claim about the inhuman proportions of a Barbie doll). He writes of various aesthetic "spectrums" or "spaces," for example "posture space," "color space," "motion space," or the "male–female spectrum." Each space or spectrum has dimensionality, and when an artist strives to tease out an essential impression or feeling in a work, they push and pull the work's features out of "normal" ranges into "supernormal" territory. It is assumed that the artist does this in order to elicit responses by appealing to peak shifts, which corresponds to what Ramachandran sees as the "purpose of art," which is to "enhance, transcend, or indeed even to distort reality."[32]

Ramachandran insists this is a deep truth about the function and meaning of art. But, as provocatively noted in John Hyman's rebuttal, Ramachandran's proposal amounts to little more than a "theory about why men are attracted to women with big breasts."[33] Ramachandran's musings, in the final analysis, fail to enlighten the question of art and its place within aesthetics. It seems rather obvious that the progressive shifting of peak stimuli is an important element in the unfolding of aesthetics, but we should also remember that all aesthetic experiences, not just artistic ones, necessarily stretch and skew our relational attractors within the field of experience. Indeed, such evolutionary explanations are so coarse-grained that, as David Buller notes, they are "wholly uninformative about the selection pressures that act on a species."[34] Though we can safely suppose that our human ancestors, like all animals, faced selective pressure to attract mates in order to forward their genes to subsequent generations, this tells us nothing about what those specific pressures were. "Simply knowing that Pleistocene humans needed to attract mates doesn't inform us on the subproblems that constituted that adaptive problem for Pleistocene humans. And it is those more specific subproblems that adaptations would have evolved to solve."[35]

But it is likely that the specific subproblems that our sedimented tastes and implicit preferences evolved to solve will remain forever inaccessible to us. As Buller insists, it is likely that the selective pressures acting on earlier humans had a lot to do with *their own* psychology. We too easily forget that our ancient ancestors were already thinking and reflecting. And thus, as in today's culture, the psychologies of the past must have had their own hard-wirings and biases, for example, their own innate or learned ways of psychologically reacting to various mating cues from their prospective partners. No matter how far back we go, nowhere is there a "blank slate" or original empty form of hominid to be found that does not already have its own exclusive environment, filled with its own specific psychological motivators and attractors, temptations and preferences. The psychologies of early humans did not fossilize with their bones, and thus our ignorance of them is most probably insurmountable.[36]

Adaptationist interpretations of aesthetics hinge on the assumption that there were *stable* overarching adaptive problems during evolution with which our ancestors were confronted. If this were true, evolutionary aesthetics could be defended against the critique of its "coarse-grainedness", its "low resolution." But unfortunately, even if humans spent millions of years in the Pleistocene, the predominant selective pressures were likely not all that static. This is plainly obvious if we consider the perpetual intrusion of psychological and cultural competition that constantly modified and skewed the actual ways human populations reacted to those environmental selective pressures. And hence, it is unlikely that any so-called stable environmental pressure would translate into a stable incremental course of adaptation: in each given situation, no matter how stable and unchanging the selective force might have been, it would confront a constantly changing cultural and genetic Zeitgeist, varying widely across various tribes and generations. As Varella, Evan Thomas, and Eleanor Rosch stressed, adaptations "satisfice" constraints: they do satisfy them, but that does not mean they answer to them in an *optimal* way. An adaptation's response to constraints needs to be only "good enough"; that is, it need only *suffice* in order to be forwarded by evolution. And since the overarching and relatively stable selective pressure can be sufficiently satisfied by multiple different responses, knowing what the stable coarse-grain pressures were tells us very little about what may have actually transpired.

This is all compounded by the fact that feedback loops inevitably emerge in these chains of influence. Evolutionary arms races and niche construction are common in the natural world and are examples of selective feedback loops. Whenever there are multiple agents competing for the same limited resources, whether they are prospective mates or potential prey, selective pressures loop back onto themselves and escalate in the form of positive feedback. A bat evolves echolocation, which allows it to locate moths, in turn provoking the moths to evolve a capacity to hear the bats and maneuver evasively, which in turn pressures the bats to evolve better maneuverability, and so on. The cheetah evolves to be a faster runner, and hence there is pressure for its prey, the gazelle, to also evolve to be faster, which in turn pressures the cheetah to become *even faster.* The male ancestor of the bowerbird builds a beautiful bower and attracts many females, which in turn pressures other males to build more intricate bowers, and after several generations, we have a species of expert architects. Niche construction similarly implies feedback: an animal finds a new niche that changes its selective pressures, and speciation begins to occur, in turn changing the animal's traits, which can now potentially afford new niches, and so on.

These "runaway" effects occur everywhere in nature, where certain characteristics between ecological actors mutually reinforce each other and form positive feedback loops. The problem for evolutionary aesthetics and psychology is that these feedback loops are constantly interacting and influencing the external selection pressures a species must face. Arms races and niches necessarily *close themselves off* from the overarching environmental influences and take off, playing their own private games, and therefore drift along unpredictably. This leads to the proliferation of complexity and nonlinear relations between species and their environments, and thus the prospect of accurately tracing a function from a specific selection pressure to a specific adaptive response becomes ever more elusive. When a new niche or arms race becomes operative and is invested by an organism, the feedback loop itself quickly begins exerting specific influence on all the nested levels of the system: genes, avatars, species, environments, and so on. The niche reinforces the *specialness* of its particular character. In birds of paradise, one niche might favor wide black tails with white tips, hopping dances, and twig architectures; a parallel niche might reinforce long yellow tails folding forward, horizontally swaying dances, or leaf-based flooring constructions. Many niches will naturally exist in parallel and immediately begin reinforcing their particular patterns. They are local *top-down* selectors, where the competition between individuals drives a progressively intensifying race to grow longer plumage or devise more complex songs and dances.

Finally, yet another problem looming on such evolutionary accounts is that of *our own* feedback loops with the world,[37] the main tenet of second-order-systems thinking. We contemporary humans also exist and evolve within niches, arms races, and other resonant, trait-reinforcing causal loops. There seems to be an epistemological block on our capacity to step out of our own bounded rationality and our implicit cognitive biases (which are incidentally also evolutionary hand-downs from our ancestral legacy) that skew our own capacities as observers of evolution, a situation compounded by the biases of our specific tools of measurement and our models of understanding. Like our ancestors, we behave in ways that are conditioned not only by overarching and relatively stable environmental selective pressure but also, and *perhaps more directly,* by our own resonances with the mutually dependent ecological actors we interact with and a plethora of nested selection mechanisms. Hence our attempts to map the relations of cause and effect onto the features of an environment and the traits of an organism are themselves constrained by the observational biases our cultural and psychological embedding incurs: we tend to see only what our biases allow us to see, what is easily understood, satisfies our inferential models, and is relevant to the goings-on of our own anthropocentric niche.

Indeed, the ecological niche and evolutionary arms race can be understood synecdochically with a notion in physics: *resonance.* The resonator reinforces oscillations of specific frequencies. In effect, it is a kind of selection principle: under tension, a string of a certain length will *select* and *reinforce* vibrations of that specific wavelength (along with its harmonic divisions). If you feed a noise source into a closed electronic circuit, the *random electrical fluctuations* contained in the noise will be filtered out and the circuit will quickly stabilize onto a specific frequency of oscillation. The feedback loop instantiates a resonant circuit that reinforces specific characteristics in the random noise, filters out the rest, selecting only the frequencies at which it can resonate. This constitutes a relaxation process for the circuit. The infinite individual variances contained in the original white noise must eventually converge upon a local equilibrium, the resonant frequency of the circuit. In evolutionary biology, the niche behaves in much the same way. In the art world, the markets, exhibition networks, and institutional frameworks have the same effect. They form resonant circuits. The relationship between the organism, environment, and the emergent niche milieu is self-perpetuating, just as is the relation between the artist's output and the patron's aesthetic preferences. The sensory-motor system of the species enters into resonances with its milieu, which, over generations, reinforces the patterns of the organism's repetitive behaviors, in turn reinforcing the colorful feathers and dances of the bird of paradise and the architectures of the bowerbird. The artist produces an aesthetic artifact and faces critique from the art world, either discursively or in practice (access to opportunities); the artist readjusts with regard to this critique; and the process positively expands with each cycle, filtering out what are locally considered unacceptable excesses and inconsistencies.

Such "reinforcements" through positive feedback or constructive interference, in both nature and culture, emerge in regions of logical and causal underdetermination. If, as we saw in chapter 1, aesthesis concerns the transition from indistinct to distinct, from non-sense to sense, from underdetermined to determined, the same can be said about how a self-sustaining and self-amplifying niche emerges from a region of underdetermination, where many influences cancel each other out. The failure of the adaptationist method of "reverse engineering" is analogous to logical positivism's failures to reduce individual statements to given stable experiences. Willard Van Orman Quine rightly noticed that beliefs are underdetermined by individual events, and similarly, the holistic character of evolution (symbiosis, genetic drifts, nested selective basins) precludes phenotypic reduction to specific selective pressures. Just as analytic truths smuggle in their analyticity through the "sleight of hand" of synonymy, our adaptationist explanations traffic in just-so stories and uncontroversial centers.

Of course, no one is denying the importance of selection and adaptation. Rather, the point here is that there is no way to confidently attribute specific selection pressures to any of the features of human behavior we call art. It comes as no surprise that certain normative aesthetic preferences fall into such localized resonant niches, which select a minor subset from the great variety of aesthetic impulses. This explains the evolution of styles and aesthetic schemata in culture, as well as the resonant properties of institutions such as museums and other venues, funding bodies, academic establishments, and publications. Each is a niche. Each social playing field an artist can invest carries out its own selective procedures on the artist's aesthetic individuation process. According to Niklas Luhmann, the coupling of artistic processes with social systems leads to self-reinforcing resonant behavior: "Communication through art tends toward system formation and eventually differentiates a social system of art."[38] And at the same time, the institutions and their constraints are also reciprocally presupposed. Even the most stable institutions hold together only because they are continually enacted by the agents who collectively take part in them (this is one of Cornelius Castoriadis's insights: the institution is *imaginary*). Once they are no longer purely "outsiders," artists cannot be said to be "autonomous" creators, for they are constrained by the norms they confront; their work is filtered by the sieve-like interface they instantiate with posterity. "One doesn't bite the hand of one's host,"[39] and the institutions implementing these historical filters hold to little more than the competing "uncontroversial centers" of the actors who continue to sanction them. Over generations, the many artistic niches populating the aesthetic manifold have been delimited and territorialized as autonomous social systems. It is this reification that gives each art institution and each competitive playing field its own resonance. New aesthetic impulses are thus pulled up into the resonant attractors of the social system of art, each attractor implementing its own selection, teasing out only what it is attuned to and rejecting the rest. So, when artists select from among their own inspirations those that have the most "social value," "relevance," "originality" or "merit" in order to pursue only those that might have a chance of succeeding in the competitive game of their art careers, their own critical faculties fail to belie the selections that implicitly operate through their gestures. There is perhaps no purely subjective or personal aesthetic "preference"; even the most estranged outsider is always already evolving within a field of selecting mechanisms. Throughout the aesthetic manifold, even before the artwork appears in the public world, there are constraints acting on the local creative impulse, and these constraints are themselves made up of previously settled or sedimented creative impulses that have been reified as niches and attractors.

However, by focusing only on the selecting pressures, we miss the real process of *aesthesis,* which is not simply the product of selections, but rather the process that leads from underdetermined, indistinct, free-floating differences to fixed and determining selective principles and judgments. As we have seen, this is why the prospect of circumventing the individual's aesthetic selection was so appealing to the avant-garde of the last century: by allowing chance or stochasticity to rule over the authorship of aesthetic endeavors, it implied that these hidden selection principles, which controlled and constrained the aesthetic field in each of its niches, could be dissolved by the divine randomness of the contingent event. It is a question of reestablishing the neutralizations of these selecting factors so that new trajectories may emerge from their reshuffling, from their asymptotic convergence upon local symmetries and compromises, wherein the possibilities for change are enfolded.

The true question is an epistemaesthetic one: how do the selective pressures emerge in the first place? The method of reverse engineering simply cannot get past very coarse-grained linkings of broad selection pressures and very general adaptations, and any more specific descriptions fall into the category of speculative guesswork and just-so story-telling, and as such, merely expresses in evolutionary terms the same problems encountered in the foundations of logic: they are confronted either with the infinite regress of premises or with a mischievous smuggling in of an unfounded justification. Succinctly put, the adaptationist reading of aesthetics puts the cart before the horse: instead of trying to figure out why we have aesthetic preferences based on prior selective pressures, we should be asking how selective pressures emerge from aesthesis to begin with. For aesthesis *is* the process that leads to these determining factors, not what derives from them. Epistemaesthetics frames the question thusly: how do selections emerge from a context of nonselection? To answer this, a drastic shift of perspective is required.

It is encouraging to now find such a shift in perspective within the field of evolutionary aesthetics itself. A recent theory of the origin of aesthetic variety in nature is advanced by Richard Prum.[40] He presents an alternative to the view that aesthetics and sexual selection take a back seat to technics and environmental selection. Specifically, he defends a return to the orthodox Darwinian view that sexual selection is actually independent of natural selection, a claim that, as we have seen, was influentially rejected by Wallace. Prum's view also picks up Ronald Fisher's elaboration of the mechanism of "Fisherian runaway." Fisher described how extreme traits and ornamentation in the males of a bird species can *coevolve* with the sexual preferences of the females of the species. Through mutual reinforcements between sexual markers and sexual preferences, a runaway process ensues, due to positive feedback, as long as the environmental pressures allow for it. Following these insights, Prum has argued that the way to understand the variety of aesthetic productions and preferences we observe in biology, as well as in human culture, is to see them as generally driven by an autonomous selection mechanism in which preferences and markers produce runaway effects through mutual reinforcement. This happens in the region of the statistical *null model,* as proposed by Fisher. As we have seen, there is a statistical chance that each phenomenon we observe is not in any way causally linked to the model we take to describe the evolutionary context it evolved to satisfy. For instance, there may be a sliding scale of degrees between the complete determination of a phenotype by a proposed genotype, on the one hand, and the absolute dislocation of the two, on the other. This latter region of *underdetermination,* where the hypothesis of a causal link— between, for example, environmental-pressure-selecting aesthetic preferences in a female bird and the extravagant plumage we observe in the males of the species—does not significantly contradict the null model and leaves the door open for a completely different form of selection that is *autonomous*, dislocated from the influences of the environmental context.[41]

Prum suggests that aesthetic preferences do not get determined by environmental pressures: as long as these are satisfied, in situations where all other pressures level out, random symmetry breaking occurs, resulting in drift. Flipping evolutionary aesthetics on its head, in this interpretation, aesthetics does not derive from sexual selection, but rather sexual selection is considered a subset of *aesthetic selection*. It is a runaway process that occurs through spontaneous symmetry breaking in regions where multiple pressures nullify each other, multiple interpretations cancel each other out, and is therefore not directly determined by any individual pressures of technical necessity or survival. It is an emergent selection mechanism that "lifts off," as it were, bootstraps itself from its substrate, and achieves independence. On this basis we can begin to speculate that aesthesic processes take hold specifically in regions where there is a neutralization of causal drives and determining selections.

This is the deeper reason that Leibniz was correct to link aesthetics with the indistinct: the process of aesthesis is initiated spontaneously in those regions of underdetermination, of vagueness, of indistinction, of undecidability. The point to take away is that aesthetic pursuits are not to be characterized as merely *useless* in comparison to technical or survival-related endeavors. The distinction rather has to do with the question of underdetermination versus overdetermination. While the technical can be seen as an overdetermination of environmental constraints (echoed again in the adaptationist's overdetermination of the coarse-grained evidence in their method of "reverse engineering"), the aesthetic essentially concerns their underdetermination. This indeed corresponds to what we considered at the end of chapter 1: even under the most obliging of constraints, artists must find the means to create in the nullifying of the determinations acting on them. Aesthetic practices reflect the spontaneity that can be made available only in these null regions, these contexts of underdetermination, which is why the creative act always begins by setting up some symmetrical distribution of potentials that is essential for the novelty or originality to express itself, such as a procedure for randomizing or for scrambling the origin, or an operation for neutralizing specific causal determinants so that the event *collapses* their superposition.

Animal Territories and Technological Enframing

There has been a related tendency in the account of aesthetic origins to associate art with the territorial animal and to understand human aesthetics as extensions of animal territorializations. Étienne Souriau, for example, asserted that spider webs and termite mounds were already works of art.[42] Deleuze and Guattari also make this link, speculating that art begins with "the first person to set out a boundary stone, or to make a mark."[43] They locate the origin of art in the establishment of its bounding frame, which is the precondition of its composition and its relation to the other than itself. They claim, "art begins with the animal, at least with the animal that carves out a territory and constructs a house (both are correlative, or even one and the same, in what is called a habitat)."[44] Following suit, Elizabeth Grosz has also endorsed the idea that art emerges "from the moment there is sexual selection, from the moment there are two sexes that attract each other's interest and taste through visual auditory, olfactory, tactile, and gustatory sensations."[45]

But how exactly are such territorializations and sexual preludes to be construed as art, or even precursors of art? What warrant's our assumption that art is in any way derivative of the territory? For, though it is true that it is difficult to imagine an artwork without some kind of frame, and that the territory of the animal is also in some sense a framing of its environment, the abstract notion of framing can just as well be seen as the general condition of all *objecthood*. And as I have already argued, if we want to avoid reinforcing the cliché that the artistic endeavor is ostentatious, luxurious, and frivolous, we should abstain from facile associations of the distinction between art and *technē* with the distinction between sexual selection and environmental selection.

Part of the allure in linking art to sexual selection and to territoriality seems to lie in the formal similarities we observe between the arts and animal behaviors, which look to us like dances and sound like music, or evoke sculpture and architecture. But the question is: how are those territorial activities that remind us of the human arts different from those territorial activities that do not? Certainly, not all territorial activities support this analogy with the arts. Souriau makes a categorical distinction, for example, between the animal that creates its territory by secretion, such as the mollusk with its shell, and the animal that externalizes its territory as "artifice." For Souriau, the key factor is this exteriorizing principle. But again, if this is true, how are we to distinguish "artistic exteriorization" from the exteriorization that, as Bernard Stiegler has observed, characterizes all *technē*? And in what way is the territory a "making expressive" of the earth, as Deleuze and Guattari claim? On the contrary, there is reason to believe that, as a prosthetic externalization of the organism's bodily function, the territory rapidly becomes "implicit" to the animal in the same way human technologies become implicit to our ongoing activities. As Heidegger showed, rather than express or explicitate, technology dissimulates or *implicitates* itself in its characteristic "readiness-to-hand." How, then, is a territory different from technology? Is technology not always a territorialization of the world?

An analogous dilemma also shows up in the theory of "aesthetic selection," which we visited briefly in the previous section. Prum suggests that art begins the moment a genetic feature or behavioral pattern is offered as an object of sensual evaluation: the reason why insect-pollinated flowers are "beautiful" while wind-pollinated flowers and plant roots "are not," he argues, is that roots and wind-pollinated plants are not offered up for aesthetic evaluation by other aesthetic agents. By contrast, the insect-pollinated flower evolves to be "beautiful" because it caters to the aesthetic preferences of its pollinators. Though this makes a certain amount of intuitive sense, it is not clear how to strictly distinguish an evolutionary feedback loop having to do with sensual appreciation or evaluation from an evolutionary feedback loop that has to do with other forms of selection. The variety of different plant roots attests to this. Different plants evolve visibly different strategies for coping with different types of soil and other environmental factors, which are in this sense no different from the insect sensory systems that select for more beautiful flowers, and by this logic, rocky soils implement aesthetic judgments for roots just as wasps implement them for orchids. Ultimately, our decision that flowers are normatively beautiful, whereas roots are not, is a weak claim that says more about our own aesthetic biases than the origins of art. In order to make it work, we would have to append all sorts of clauses: does the aesthetic selection require a biological organism doing the selecting, sensory organs, a nervous system, a brain?

More generally, though, if we think of art as an extension of the "imperative to organize the space of the earth,"[46] it is difficult to see how this activity of "framing," on its own, is any different from the concept of *Gestell* (literally "framing"), which Heidegger understood as the essence of technology. By examining how art can be thought of as framing while nevertheless being different from the sort of framing implied by *Gestell*, we might gain some further purchase on the question of the art–*technē* distinction.

We can situate this question within the material context as a compromise between organism and world. For Simondon, the early human tribal life-space was varied from region to region, being populated with *key points, peaks,* and *thresholds,* in a field of varying degrees of intensity. Similarly, for Deleuze and Guattari, the *presignifying* life-world is directly connected to the natural flows of the earth. The human species and culture is inseparable from its milieu, with flows and pulses of vegetation, animal herds, weather patterns, seasons, tides, semen, and excrement. The rhythmic motions of the earth, moon, and sun synchronize and modulate our endogenous circadian clockwork. These periodic phenomena always already compose the human life-world as a "milieu," and the intersections or the boundary crossings between these milieus are home to activities of what Deleuze and Guattari call *transcoding,* the source of the characteristic rhythms of the life-world as experienced by early human culture.[47] Consequently, the presignifying world of the early human is already an assemblage of what we might call *partial objects.* They are still concrete and topologically mixed. The boundaries between them have not yet been discriminated; their various regions of activity have not yet been abstracted from the mix and discretized as individualities. No mark has been attributed to them, and so their limits are indistinct. Signification and reference appear only once the peaks in these intensities are framed, territorialized, named, or measured—that is, when these partial objects are severed from their contexts and a "completion" or term is imposed with attributes that now belong irrevocably to them. Indeed, "having" attributes and properties implies that a selection has taken place. Objecthood, as such, is an enframing of chaos. Only once we have entered this regime does numbering and indexing become possible. The transition from indistinction to distinction involves a territorialization and a discretization of the world. The "territory is the product of a territorialization of milieus and rhythms."[48] But what exactly is the nature of the change that takes place between the natural state of these rhythms and flows, on one side, and their territorialization, on the other?

Deleuze and Guattari suggest that territorialization has two aspects: on the one hand, it leads to the specialization of tasks, occupations, trades, and professions, and on the other, to the emergence of rites and religions. Simondon previously put forward a similar claim in his speculative anthropology: psychosocial individuation proceeds from an original division between technology and spirituality. In some sense, then, the technical and the artistic are two modalities of territoriality. Here we get a glimpse of how the interplay of aesthesia, art, and *technē* participates in the question of territory. There is "no reason to conclude that art in itself does not exist [in occupations, trades, specialized tasks], for it is present in the territorializing factor that is the necessary condition for the emergence of the work-function."[49] In other words, there is an artistic inclination that subsists within the emergence of the territory, perhaps as part of the mechanism of emergence itself. We might conjecture, provisionally, that aesthesis is that aspect *in* territorialization that makes the earth and its flows expressive, rather than the aspect that submits them to implicitation.

So we must remember that Deleuze and Guattari find an essentially positive aspect of territorialization: the act of framing the rhythms of the world and offering them to sensation, making the functional expressive. It is not founded in the *negation* of the other, which is sometimes how territoriality is discussed in other contexts. For example, they explicitly disagree with ethologist Konrad Lorenz on this point.[50] The expressive aspect of territorialization is thus a positive notion, an inclusive process, distinguished from the settled territory, which is a nonexpressive implicitation and technical instrumentalization of the environment. So, there is a sense in which this specifically expressive character concerns a turning of technical instrumentalization against itself. Technology as *Gestell* is already a framing of chaos, but it does not make the chaos speak for itself or stand for itself. Rather *technē* is characterized by a process of *implicitation*: it becomes implicit to the ongoing deferral of sensation, while it submits to chaos in the process of deferral. There is a kind of suspension of technical deferral in the initial inclusion of the rhythms of the world into the territory's form of expression. To the extent that art emerges from intensifications of the real by explicitly aligning sensation within chaos, by monumentalizing and intensifying chaos itself, by getting the implicitations of *technē* to counteract each other so that no deferral of sensation takes place and so that sensation is allowed to bathe in the ambiguous chaotic element, we can say that art specifically concerns the cusps and thresholds *between* these technical enframings of being, which are, by contrast, takings-for-granted of chaos. *Technē*, in other words, *is* implicitation: it dissimulates itself in its process of deferral. The artwork, by contrast, is not *technē* itself, but something *within* it, not territorilization itself but a potential aspect of the territorial imperative, not framing itself, but an engagement in the ambiguous and ambivalent inclusive principle that is the condition of the frame's exclusionary discrimination. Art is an intensification of this character in framing that escapes the organism's functional processes of self-preservation, extended through the territory, whether it be an architecture or a secreted shell.

This is precisely the point made by Jean-Marie Schaeffer in his critique of the reduction of art to animal sexual selection. He contributes a distinction between the *depragmatized* aesthetic moment of, for example, the precoital parades of male bowerbirds and the *repragmatized* functional relation they result in.

> The conclusion of the parade, in this case the fact that the female connects the attractivity of the signals to the emitter of these signals and (eventually) mates with him, is not part of the parade's structural homology with the aesthetic relation. It falls under the specific function fulfilled by the depragmatized attention, which, in the case of the bowerbird, is the function of sexual selection. This function, which could not be more pragmatic, garnishes the entire depragmatized ritual interaction.[51]

Schaeffer thus mounts a case against associating the functional role of the territory with the intrinsically defunctionalized, depragmatized process of aesthetic intensification. Despite its eventual refunctionalization, reterritorialization, or appropriation by technical processes, art can thus be thought of as a specific kind of proto-*technē* that provisionally turns the technical or territorial function against itself: something in the play of deterritorialization and reterritorialization that suspends the function of extending the organism's phase space (or territory). If Kant was right to identify a *purposeful purposelessness* in art, it is because the artwork is expressive by virtue of its *refusal* to be reduced to the functions of the tool and the implicitation of the territory, and hence maintains a paradoxical status. Grosz astutely notes that, though "framing creates the very condition for the plane of composition and thus of any particular works of art, art itself is equally a project that disjars, distends, and transforms frames, that focuses on the intervals and conjunctions between frames." An artwork is an enframing of chaos, but it is the tension it creates with its frame and with framing in general that makes it art rather than *technē*. This is indeed why successful artworks conserve an untimely ambiguity or undecidability, and it is this that constitutes their "monumentality," that allows them to "transcend the age": they eschew their reduction to *any individual* function, reason, or purpose. The epistemaesthetic relation is a logical precondition of the initial marking of the territory, the judgment, or the measurement. It hovers over the territory, beyond it, in the *utopia* off the map, within the intervals of our measurements, or beyond our reductive capacities of discernment, which is why it remains intrinsically indistinct. For, if the territory "bites into [milieus], seizes them bodily,"[52] the milieus and regions rendered expressive in their aesthetic intensification are only *secondarily* integrated into the territorial boundary. Territorialization accounts for the modeling, indexing, or referencing of the milieus and rhythms it selects and captures.

These two moments, the expression and the implicitation, mirror the entangled levels of cognition: selective judgments and framed discernments may naively be said to concern the "higher" cognitive faculty, which cannot be fully parsed out from the indistinct and sensuous conditionals of the "lower" faculty. Furthermore, the territory is inseparable from the epistemeasthetic production of intelligibility that we explored in chapter 1: it makes the world make sense, filling it with meaning. In that case, echoing the language of topology, since one cannot smoothly or continuously deform one manifold into another, we might modify Grosz's claim slightly: disjaring and distending do not seem to be sufficient to bring about world-model collapse. The territory must truly be *transformed*, break, discontinuously leap from one incommensurable territorial model to another, and actually change the topology of the system, the routings of the ins and outs of the organism's relation to the environment, the ordering principles that legitimize its engagement with the world. For, the aesthetic practice is not just a distorting of frames; it involves the intensification of the catastrophic event of the territory's collapse.

Such catastrophes can be likened to the "liminal spaces" described by Victor Turner, who observed that rituals explicitly concern *limit crossings*. Rites of passage, for instance, are aesthetic intensifications of the crossing from childhood to adulthood, a crossing of a symmetry-breaking threshold. The previous frame or territory is collapsed and replaced by a new one, but in the transition, there is a zone of ambiguity, a suspension of territoriality. In between the territory of youth and the territory of adulthood, there is a liminal space that characterizes the ritual itself: it is a place outside of space, a moment outside of time.[53] With the rite of passage, what would be a continuous sequence of maturation is cut or broken and marked by the "inter-structurality" that it instantiates at the site of the catastrophe. During the ritual, after it begins and before it ends, you do not have any status, any age; you do not even have a name. So what exactly is it that is being cut or broken? It is not the milieus themselves, nor the rhythm that characterizes the transductions between them. Rather, it is the territory that codes them or frames them. It not only deterritorializes the territory of the child but also keeps the catastrophe open, or stabilizes the event of rupture. It marks the deterritorialization and attempts to maintain its essentially open, broken, liminal state. Without the ritual, the crossing over to adulthood and the deterritorialization it implies might happen anyway. But by maintaining the open rupture so that it "stands for itself," like a monument, the ritual or the aesthetic intensification of a deterritorialization not only transitions from one world model to another, but marks or inscribes the event of transition into the consequent world model.

Like earlier cosmogonic or religious paradigms, scientific paradigms, which Thomas Kuhn argued advance in radical shifts or *breaks,* also structure our implicit views of reality. But given that it presupposes a transition, a transformation, the paradigm shift implies the crossing of an ambiguous limit that cannot be said to belong to either paradigm. Aesthetic practices, then, can be said to occupy this nonterritory, this nonparadigmatic in-between, and often involve an intensification of the topological rupture of the world model or model of objectivity. If the artwork captivates us, it is precisely by leading us astray from the otherwise settled field of invariances that provisionally structures intelligibility. It intensifies the moment the world collapses. It delays and textures the infinitesimal lapse between one ordering and another. It is a proliferation of mediations between territories that synergistically counterbalances their incommensurable causal expansions. The work of art is the production of a metastabilized collapse of orders and invariances. Hence, art is not so much a specific method of contributing to the *partitioning* of the fabric of sensation, as in Rancière's political-aesthetic philosophy, nor specifically the act of *territorializing* the milieus and rhythms of the earth. It can be better described as a *care* or *concern* for the symmetry-breaking event, for the paradoxical continuous catastrophe implied by the movement from one paradigm, cosmogony, or model of objectivity to another; it is a cultivation of the *u-topia* between *topoi.*

The artwork, as a historical memorial or marker, is a knot in the field of individuation, a tunneling event, a passageway to a new topology of invariances. It sweeps participants away in a new transduction of the world. As Stiegler notes, there is often a "mystagogical" aspect in the artwork. [54] There is indeed an abductive element in art that seeks to persuade while nevertheless avoiding charlatanry, which for its part would amount to a hijacking of the *relativism* of the artwork's reception, an insistence on a privileged perspective, an officializing interpretation or judgment. But this decision to take part or not to take part is less a mere subjective act than a real condition of the situated relation between subject and object, between organism and world. For, the mystagogical element in the artwork marshals precisely the same *faith* that allows for the fabric of objectivity to hold true, the same suspension of disbelief on which each act of inductive prediction relies, and without which we would all solipsistically doubt that anything before us was indeed the case. The continued apparent stability of causality, as Hume showed, is just as suspect as the gambler's superstitious optimism that his improbable winning streak will persist. But art turns this causal superstition against itself, producing a paradoxical faith in the collapse of the model of objectivity. This is why aesthetics cannot be reduced to matters of judgment, for by contrast, judgment "prevents the emergence of any new mode of existence."[55] From a safe distance, the artwork is a strange beyond, a black hole from which no information can escape, from which we can gain no insight. But, from such a distance, one is not "moved" by the artwork; one does not participate in its intensification of the trans-paradigmatic limit. The artwork's event horizon must be breached, the technical impulse submitted to the cruelty of its ambivalence, in order for one to live or to cognize the topological catastrophe, the territorial collapse that the artwork paradoxically enframes. The artwork, like the rite of passage, explicitly positions those who take part within this interval, this gap, this zone of indistinction between the models of objectivity that structure our world.

If these considerations are at all correct, it is misguided to simply equate the emergence of territory with the emergence of art, or as in the case of Souriau, to interpret the "beauty" of a butterfly's wings, a spider's web, or the typical dance of a rooster's sexual prelude as forms of animal art. It is not the territorial itself, but rather a potential *in* the play of deterritorialization and reterritorialization, that reverses the priority of preservation; it is not self-organization itself, but rather something *in* the self-organization that bootstraps variation in a seeking out of new potential intercessions and supersessions. Though aesthesis may terminate in territories and judgments, it does so only once the gap it instantiates is closed in its wake. To *believe* in a certain model of objectivity, to give oneself completely to the *Umwelt,* is to uncritically accept or trust the a priori conditions of a given paradigm of cognition. It is in this sense that the technical or territorial impulse is a suspension of disbelief, a superstitious leap of faith. Art, on the other hand, mobilizes a *Pyrrhonistically* rigorous reason: it is critical of the territory, and thus implies a *suspension of judgment* on the model of objectivity, an *epoché*. Intrinsically, as aesthetic intensifications of boundary crossings, artworks are less concerned with creating territories than with opening and maintaining nonterritories, widening the gaps and fissures between territories, revealing otherwise hidden horizons for the subject–object correlation, new speculative transcendental *types* of cognition. As Theodor Adorno observed, the aesthetic shudder is a desubjectivizing event: artworks deterritorialize the judgments of those who truly submit to them. This is perhaps what Deleuze and Guattari fail to adequately stress. If art is that which conserves itself, as they claim, it is a conservation of the internal difference from self that constitutes the *milieu of all milieus,* the chaos that surrounds and infects all territories at the limit. Art is an *a-territorial* expression as long as its essential breach is maintained.

Artworks generally retain an element of this a-territorial liminality, even though this might not be observable from every frame of reference. The artist attempts to keep this breach open, as a piece of chaos that stands for itself.[56] It is in the gap that creativity necessarily takes hold, for here chaos is allowed to flow into and disturb our networks of reference and constraint, our *total science.* The territory is indeed, on the one hand, a "poster or placard," a signpost that announces or expresses a given milieu. But, on the other hand, it is also immediately the *making implicit* of the distinctions and variables captured by territorialization in its prosthetic supplementation of the organism's "internal" milieu. The territory of one animal may be expressive for the other animal, but from the point of view of the animal within its own milieu, the territory becomes implicit to its activities of subsistence. Hence, intrinsically, for the animal that constructs it, the territory is analogous to the tool or technical object, and as such, takes part in the same process of implicitation that is the essence of habituation, the movement of learning-forgetting and mnemotechnical exteriorization. This implicitation undermines the expressive and the initial creative gesture.

By contrast, in its purposelessness, art is *intrinsically* a-territorial, while being territorial only *extrinsically,* in its eventual intersection with various territories. It is hence only extrinsically that the practice of art is a "waste" of time or resources when compared to the "useful" activities we relate to technics. The *chaoïd variety* that, according to Deleuze and

Guattari, characterizes the artwork is unthinkable without an internal rejection or avoidance of the implicitation sustained by the organism staking a territorial claim in an act of mnemotechnical supplementation. The work of art attempts to avoid capture by this implicitation. It is not just *sub-lime* but truly *liminal.* Artworks emerge from the *limit* of sense. Even in those cases where artists capture or express the topologies and territories implicit to experience, they do so by explicitly opening a gap within them, and only for as long as the gap is maintained. The artist is a funambulist, the artwork a house of cards, a balancing of inertias and tendencies, an epochal suspension of judgment. This, in sum, is why the *chaoïd* object is inherently aterritorial; its consistency is that of a *difference from self,* which is maintained only as long as it eschews its absorption into the territory or resists its submission to the given model of objectivity.

3 AESTHESIS AND PROSTHESIS

PROSTHESIS =
TECHNICAL
EXTERIORIZATION [MER]!
(and technology)

Aesthesis as Integration

It is time to combine some of our insights from chapters 1 and 2. We will now turn to how aesthesis and *prosthesis* (technical exteriorization) relate to matters of memory and cognition. There are several steps to take here. First we must come to consider aesthesis as integration, as synthesis. We will see that the scope of such aesthesic integrations necessarily exceeds the limits of human cognition, and indeed must be attributed to all biological organisms that are defined by their business of grouping events together, drawing boundaries of inclusion and exclusion within which the many become one as a collection or a set, and of ordering the before and the after in succession. Once we have resolved a general nonhuman sense of integration, we will be in a position to compare the results to better known accounts of the experience of time, which is, as philosophers since Augustine have observed, inseparable from memory. As much as possible, we should strive to give a naturalistic account of these transcendental conditions of aesthesic integration: we must admit that these retentions and protentions, these syntheses that define the diachronic limits of each moment of experience, are somehow inscribed and expressed in flesh, in matter. Along the way, we will come to better resolve how exactly technical and artistic exteriorization fits into the greater nonhuman landscape. I have already somewhat underhandedly evoked what I have come to believe is the relation between aesthesis and prosthesis, but we will now circumscribe the intricacies, the ins and outs of the mnemonic organism in its essential striving for life, which already, in its very functional, processual structure, suggests the prototypes of

TIME:
memory

MATTER

THIS!!!

what we humans have come to know as art and technology, two variants of human exteriorization.

Our starting point is Gilbert Simondon's obscure theory of aesthetics, which, instead of viewing aesthetics through the lens of *judgment*, as did Kant, suggests that it rather concerns a principle of abstract togetherness or *integration*. Simondon proposed to see aesthetic thought as a reticulating, unifying tendency between the separate individuations of human culture, rather than as an individuation in its own right (such as a product or by-product of adaptation).[1] This view of aesthetics follows from his speculative historical anthropology, which traces a genealogy of *technē* through its successive stages, each defined by resolutions of local tensions through discontinuous innovative breakthroughs and continuous efficiency improvements. Human technosocial individuation had to originate in a kind of "magical" paradigm of early human cultures in which the seeds of what would later become the offshoots and tangents of culture were suspended in an undifferentiated mixture. Progressively, he reasons, the technical would have disentangled itself from the mythical and the spiritual dimension of early culture—what he misleadingly refers to as "religion"—and become its own autonomous field of thought, taking off on its own and leaving the "religious" in its wake, a transcendental background for the immanent foreground of the technical. Like the force of gravity in the universe, in this view, aesthetic thought pulls together, ever so slightly, the branches of human technospiritual individuation.

> Figure and ground separate by becoming detached from the universe to which they adhered. The key points become objective, retaining only their functional mediatory characteristics, becoming instrumental, *mobile*, capable of efficiency in any place and at any time; as figure, the key-points detached from the ground for which they were the key, become technical objects, *transportable* and *abstracted* from the milieu.[2]

Simondon's discussion of how human subjectivity and technics emerged from a preconscious confusion echoes his unorthodox philosophy of information, which describes a material inertia or tendency toward higher order and complexity. He imagines the prehuman lifeworld to have emerged in successive nested levels of differentiation, much like the far-from-equilibrium thermodynamic systems studied notably by Ilya Prigogine, where order emerges from randomness. Simondon starts with one major event of *disparation* between figure and ground, with the world dissected into two halves, one that was *objective* and led to technical, functional, instrumental individuations, and the other *subjective*, leading to moral and psychological individuations. He insists on the materiality of this process and claims that it corresponds to a transition from an undefined continuity or unified whole to a set of individualized, discretely distinguishable individuals.

As the network of connections between the "key-points" of the primitive world is ruptured, on Simondon's account, its characteristics are freed and no longer tied to specific places or times and now "hover over the whole universe, throughout space and throughout time, in the form of detached powers and forces." This speculative anthropology foreshadowed how, in Gilles Deleuze and Félix Guattari, *territorialization* was said to result in both the specialization of tasks and the dogmatization of spirituality. It is this division of the territory into its two aspects that provokes the internal thrust for the advancement of both *technē* and spiritual thought, as the first discrepancy in the context, the first distinction, or the first *mark*, initiates a process that propagates through its state-space. In Simondon's preferred example, the crystal propagates through a supersaturated solution, from point to point, aligning the molecules according to its symmetrical organization and the constraint of thermodynamic efficiency. In the same way, the territory's liberation of the task, of the technical and inferential, creates a new propagation. An abstract process of discretization now propagates through the world, sampling continuous intensities from place to place, detaching the peaks from the troughs, and forming countable objects or atoms. It transforms (disguises) the partial object into (as) an object, wresting the individuating process from its pre-individuality and submitting it to its individualized finality.

In an extension of Simondon's insights, we can give a general account of the concept of *discretization* as the character of any process that transforms continuous functions into discontinuous ones. By producing sets of mobile atoms, by everywhere excluding the middle, and by ridding the system of indistinction and generating a series of discrete states with no in-betweens, discretization allows for the proper countability, namability, and orderability of the world. Discretization populates the world with discrete things, rigid and passive segments, rather than active partial objects or processes. Objects are now free to combine and constrain each other in ways previously impossible, since they now have their own positive autonomy, no longer tethered to the surrounding flows. It is thus the underlying condition of computability and of the digital domain. The development of *virtual computing* (which implies Turing universality) happens at the site of the decoding of one milieu by another, extending biological nature's own propensity to virtualize itself in previously hidden fractal dimensions of reality. Discretization is the process by which, in the canonical engineering example, an *analog–digital converter* measures a continuous signal: if the signal potential reaches the converter's specific threshold, it will sample 1; if it does not, it will sample 0. But this process is already at work in nature, in between the levels of emergence, as one stratum decodes another, using it as a basis for itself. For instance, environmental selection can be thought of as such a digital sampling: an organism's behaviors are forwarded (1) or deleted (0) by evolution accordingly with its success or failure at satisfying environmental thresholds. Similarly, Derrida's reading of François Jacob led him to recognize a similar connection between *writing* and the essential *programming* of all life.[3] For, biochemistry is also a sampling of the chemical domain, where the information of life resides *untapped* (coded) before it is decoded by abiogenesis. Life is, in this sense, a decoding of the chemical realm inseparable from a discretization of the molecular, just as the human's linguistic or signifying propensities are decodings of its own animal-environment reticulations. Each level of emergence, each site where one substrate decodes another, implies its own irreducible dimension of process and procedure. With each level of emergence, a world leaps out of itself: the symmetry of the "substrate" is broken, and new affordances, attractors, and relaxation processes are put into play. A new *type* emerges, with its own limited array of potential constructions. Nature everywhere decodes itself, generalizes itself, masks this area from that, and overlooks some of its own details in order to express others more distinctly.

As Bernard Stiegler has observed, technicity is inseparable from this "discretizing" process. It results from the tendency in human history favoring the division of labor and the specialization of tasks. Stiegler's use of the term "grammatization" extends Sylvain Auroux's concept[4] and describes the process at work in making discrete and parsing out continuous sensations. The progressive segregation of various fields of human life into distinct categories and territories, following the general thrust of technogenetic discretization, is the appearance in the human world of a more abstract and general process of discretization. As we saw briefly in chapter 2, natural selection too finds it more energetically efficient to take thermodynamic shortcuts by producing specialized faculties and organs, which leads to the separation of the faculties and psychological biases.

This was known to Henri Bergson, who remarked in *Matter and Memory* that civilization's "division of labor" mirrored biological evolution's increase in complexity. The "higher" life forms are increasingly segmented into organs, and the integration and cooperation of all these distinct functional entities constitutes the organism's complexity.[5] The process of civilization has simply followed this trend toward the division of technical tasks and thought processes attested to by the remarkable variety of autonomous human activities and institutions of behavior we observe today.

Simondon's account of the origin of aesthetic thought is founded in similar reasoning: religion and technics, ethics and science, and their subsequent ramifications, emerged as discretizations of earlier instances of culture where the differences between these fields were still unexpressed and potential. History has seen the progressive separation of words and numbers, signifiers and signifieds, the grammatical ordering of subjects and predicates, and the specialization that results in divisions of labor. Not so long ago, there was great overlap among religion, philosophy, and science. The disciplines and domains of culture today are *individuations*, more exclusive or specialized offshoots of prior more inclusive domains. Simondon sees *technē* as a thermodynamic force of nature: a field of phase shifts evolving through successive resolutions of tensions, wresting partial individuations away from their localities and organizing and functionalizing the processes of the world. But, in the far past, Simondon speculates, our ancestors experienced their world as a *magical unity* containing no strict distinctions between domains.

> Primitive magical unity is the vital relation link between man and the world, defining a universe at once subjective and objective prior to any distinction between object and subject, and consequently, also prior to any emergence of the separate object.[6]

Crucially, however, instead of categorizing aesthetic thought among the many individuations of human activity that have ramified from earlier magical, mythical, primitive modes of thought, he regards it as a residual *reticulating* tendency between the distinct individuating offshoots, a specter of the unified origin.

> We suppose that technicity results from [the] phase-shift of a central, original and unique mode of being in the world, the magical mode; the phase that balances technicity is the religious mode of being. Aesthetic thinking emerges at the neutral point between technics and religion, at the moment

of the division of primitive magical unity: this is not a phase but, rather, a permanent reminder of the rupture of the unity of the magical mode of being and a search for a future unity.[7]

On his account, technology and religion are the two main limbs that have unfolded from the primitive paradigm, and for its part, aesthetics remains tied to the magical realm, like an echo of the original unity. Generally speaking, it implies that aesthetic thought never forms its own autonomous field of activity, for it *is the very relational tension* that "horizontally," as it were, articulates the "vertical" offshoots of hominization's ramification into specialized tasks. In other words, aesthesis is *orthogonal* to discretization. This generalization of Simondon's conception of aesthetics comes at the cost of his unfortunately somewhat clumsy conception of religion. We know, after all, that religion itself is a systematized set of dogmas that in many ways suggests a form of *pseudo-technicity* operating on the ethical and affective domain, offering solace to the mourning, hope to the suffering, moral guidance for the youth, and so on. If we accept that the true character of the "original" division is unknowable, contingent, or undecidable—reflecting what Stiegler calls the necessary "default of origin" that defines human technicity—and hence not necessarily the character of the scission of technology and religion, we can nevertheless retain his understanding of aesthetic experience as a general unifying tendency that backgrounds all individuation, perhaps even a promise of future unity, an abstract asymptotic process of integration, amenable to Whitehead's notion of *concrescence*.

Aesthesis is thus the process of producing this togetherness in the disparate; it is the counterpart of discretization. It may be generalized as the first moment in the mechanism of emergence: it is the *intercession* that links together the preindividual aspects of parallel individuations, preparing their eventual *supersession* (the emergent phenomenon). It is a productive neutralization that prepares the terrain for the subsequent emergent resolution. The parts need to come together, to integrate, in order for the whole to lift off from them and retroactively constrain them. Like a kind of *gravity* that pulls the disparate together toward coalescence, aesthesis behaves as a transductive countertendency to discretization. Indeed, as in the speculative physics of Schelling, the question is that of what relates one thing to another within a "sphere of affinity," an abstract tendency for the disparate to be put into relation.

key => DIFFRACTION PATTERN

> This mutual tendency of all parts toward one another is not explicable other than by a *common* tendency *of all parts* toward unification in a *third*. . . . This shared tendency to unification in a third is just the bonding power which holds all parts together. Now this third would have to be something necessarily *outside* of the mass.[8]

Like the atomists' *clinamen,* this tendency is an explanation for the otherwise unwarranted coming-together of the things of the world into macroscopic entities, into groupings of attributes. Aesthesis can in the same way be likened to the process by which the many *concresce* and become one in Whitehead, as well as the principle of *compossibility* in Leibniz: it is the mechanism by which the facts and events of this world relate to each other, however loosely, how things "hang together" in the first place, and also therefore, the principle of the bifurcation and divergence of incompossible worlds. As the "empirical constitution of matter," aesthesis is that which maintains compossibility, and as we saw in Chapter 1, that which produces through habit the relative stability of causality within the sphere of *total science,* the model of objectivity, or the fabric of intelligibility.[9]

Such a speculative paradigm offers a way of accounting for aesthesis that contrasts unambiguously with the adaptationist explanations of the aesthetic domain we considered in chapter 2. For Simondon, aesthetic intuition extends beyond the categories and the disciplines of individuation and serves as a unifying tendency, a postindividuation analogue of the magical unity of the preindividual origin, a neutral locus between individual phases of being, or a local, stratum-specific field that pulls partial individuations toward an asymptotically approximated unity. Aesthesis concerns the conjunction of the disparate and discretized through a proliferation of intermediate reciprocal connections and interactions: it is the asymptotic tendency toward continuity, articulation, and resonance. It is a hyperphysical transduction between things. It produces the consistency of the world as an a priori for change and distinction. Aesthesis is a process that is *complementary* to that of grammatization and discretization, opposed to the divisions of modal experience and the specialization or functionalization of gesture and thought into faculties and biases. Aesthetic thinking, Simondon notes, "is never a limited domain or a given species, but just a trend; it is what maintains the function of the whole."[10]

Furthermore, Simondon notes that aesthesis expresses itself materially and does not reduce to the "religious" dogma. This inability to be

Aesthesis expresses itself materially

Prostheses:

reduced can be taken to imply that, though it "maintains the function of the whole," the aesthetic field nevertheless does not consist of a monistic *completion*. For, it is suggested that the process is *asymptotic*, always incomplete. Schelling reflected similarly on his conception of the constitution of matter, noting that the process "remains a tendency after all, and never achieves unification."[11] Picking up on this scheme, we might propose that aesthesis is neither magic, nor religion, nor *technē*, but rather that it concerns the real *interstices* of the individuating offshoots and dialectical breaks that populate the world with these branches of genetic activity. Aesthesia is what is left in the wake of each division: it is a nonunity, a totality that is always incomplete but subsumes and maintains a certain conditioning influence over its individuals. The aesthetic field's power thus also resonates with Simondon's concept of *transduction*: it is the abstract principle that pulls materials together into new concretions. It is a compounding of incommensurable sensations from different genealogies into an *inconsistent multiplicity* that is neither a unity nor a totality, but asymptotically approximates the absolute whole: a perpetual renormalization or passage to the limit.

Notice that this asymptotic character of the process, if correct, could begin to explain discretization itself. Just as the future unity tended toward by aesthesis is incomplete, aesthesia is correspondingly *nontotalizable*. The nontotalizability of the world is intimately linked with discretization: the asymptotic unity's nontotalizability could well be the logical condition of the "dephasing" that gets discretization going in the first place. We may be condemned to iterate simply because there is no possibility of totalization. Simondon's view of *technē* as an ontogenetic force that ripples through the prehuman's unity with the environment, detaching key-points from their contexts, and in so doing freeing previously hidden capacities and functions, could be seen as resulting from an ontological condition of incompletion or nontotalizability of the aesthesic process, or the insatiable unrest that characterizes matter in contemporary cosmology. Whitehead was right to note that, in concrescence, the many become one but are also *increased by one*. Schelling saw that the "empirical constitution of matter" implied a *third* term, a notion of thirdness, an excess or *incompletion*. Thus, in addition to the *intercession* that unites the parts coming together in the whole, material emergence is inseparable from a process of *supersession*, a leaping-off from the substrate. If "more is different"—the title of P. W. Anderson's famous 1972 paper—it is because, for understanding large-scale phenomena, it is never enough

Discretization: the process of transforming a continuous valued variable into a discrete one by creating a set of (contiguous intervals (cut points) that spans the range of the variables values

each time Ulysses is mad

to simply extrapolate the behavior of microscopic phenomena. When we increase the quantity of parts in interaction in the "n→∞" limit, we inevitably encounter breaks in the underlying symmetry. In other words, our assumptions about the scalability of the phenomena we observe are always flawed. So much depends on the level of abstraction at which we apprehend the object at hand. The *quantity* of bodies interacting has an effect on the *quality* of the behavior observed. But their quantity is conditional to their *reciprocal inclusion* within the set they mutually compose, their transduction and affiliation in the molar whole. Hence it is impossible to study the interaction of two bodies and accurately predict what will happen when thousands of such bodies interact, since the emergent phenomenon, which holistically supersedes the parts, is necessarily dependent on the passive mutuality that defines their quantity, their togetherness as a set.

These considerations should be compared to Whitehead's comments on Zeno's paradox, where he suggested that intelligibility depends on an arbitrary halting of the regress into the infinitesimal subdivisions of the continuum: "The ultimate metaphysical truth is atomism."[12] And yet, as Gilles Châtelet observed, even the point in the lonely void is unintelligible without the virtual *spatium* it intersects. Intelligibility requires that we have not only individual tokens to manipulate but also a limit under which they can be subsumed. Intelligibility requires a horizon. We may go on counting and iterating forever and never come to see the set as a whole if we do not pass to the limit, if we fail to pass to the vanishing point. As Châtelet put it, we might see "one, two, three . . . n poplars, but [would] never manage a *row of poplars*."[13] In other words, infinity as a limit or a horizon is the condition of the finite. Without the horizon of actual infinity, there can be no tame finitude of distinctions. This indeed was one of the fallouts of Gödel's incompleteness theorems: the finite cannot be formally defined without an axiomatization of infinity.[14] The limit must come first, before the individual members can be recuperated. In the same way, emergent phenomena depend on the fusion, the con-fusion, of the distinct parts making up the provisional whole. Even a completely discretized world is not intelligible until its elements are put into relation by some principle of togetherness, according to resemblance, contiguity, symmetry, and so on. So, to Whitehead's remark that "in the present cosmic epoch there is a creation of continuity," we must respond that, in fact, the very concept of cosmos, as a principle of order and intelligibility, always presupposes that very continuity, or at the very least, its approximation at

[handwritten: Bergson]

the limit. Resonating with Bergson's critique of the possible as a mislead-
ing predetermination of the future and Deleuze's subsequent assault on
ontologies of identity, we must admit that the atomism of the real always
results from, as Whitehead himself elsewhere noted, "an arbitrary halt at
a particular set of abstractions."[15] The supersession of the parts is possible only through a prior intercession by which we go from particular
to general, micro to macro, local to global. The notion of aesthesis as an
asymptotic tendency toward unification that is aligned with these considerations on the constitution of matter as it expresses itself across scales
(which also concerns an abstract principle of integration) can be generalized and construed as a material transcendental condition of cognition,
of the constitution of subjectivity and objectivity.

Primary Integration   *[handwritten: memory]*

Cognition is necessarily bound up with the question of memory: knowledge must be retained in order to be known. An important part of cognition has to do merely with a capacity for the *retention of events*. It seems
insufficient to simply have something before the eyes: we also need to
acknowledge the object before us; it must somehow be registered; it must
make a difference for our organism. The object must impress upon us
something of its character, for if it does not, we cannot be said to be cognizing anything at all. Waves of light must refract from the object's surface and emit varying frequencies of electromagnetic disturbance to our
photon-sensitive retinas, which translate them into electrical impulses
that can be interpreted by our brain's dense network of neurons, which
become appropriately charged in response. Roughly, this is how the body
materially registers the color of the object before us. The brain materially
and functionally responds to the features of the object it interacts with,
and more generally, the entire physiological apparatus changes in consequence to variations in the environment. In other words, to cognize is for
a body, for an organism, to adopt certain states that are consequences of
changes in the state of the world.

But still this exposition does not seem sufficient. A stone rolling down
a hill is also changing in direct consequence to the varying surface it
encounters. Is it cognizing the slope? The "object-oriented ontologist"
will stop here and say that, indeed, the rock is in some sense cognizing
the hill, and the hill cognizing the rock. But the living thing is somewhat
different from a rock. It is structured in such a way as to *avoid phase*

transitions. At twelve hundred degrees, all rocks, no matter what kind, will have melted away into a glowing magma. Of course, living beings (at least those biological organisms we have encountered to date) would also have long vaporized away at that temperature. But the difference is that, as the temperature rises, living beings will perceive the approaching thermal threshold and act to ward off the fatal transition into the liquid or gaseous state. The rock would just "let itself" melt, while the organism, if it is functioning properly, will take evasive actions. The organism needs to stay within thermal bounds to keep functioning. It also needs to find nutrients from which it can extract energy for doing work and chemical compounds for reproducing its tissue structures. The only way it *knows* that the environment is getting too hot or that it is running out of nutrients is for it to be organized in such a way that these approaching thresholds cause it to respond accordingly: to sweat or to eat.

Of course, from an objective standpoint, all of the processes that come into play when the organism adjusts itself in relation to the environment are just as deterministic as the rock's reacting to the hill as it rolls down the slope. But by looking at things in this way, we overlook the "purposive" character of the organism's striving to remain within sustainable limits. The essential difference between the rock and the organism is that the organism reproduces itself; it sustains itself as an ongoing process. It is thus constrained by the environment's challenges to that sustainability, whereas the rock does not submit to such constraints. While the rock has no *concern* for what is happening to it, the organism, in some sense deterministic and causal sense, does care, for its very constitution *is* the self-preservative pursuit of continued existence.

By extension, cognition too is inseparable from this pursuit of the organism's continued existence. And of course, so is the concept of *memory*. Memory is nothing if not the basic capacity for the organism to *recognize* that the approaching threshold of sustainability is a danger or to tell whether that molecule on its membrane is food or poison. A cell membrane is already a mnemonic structure: it materially constitutes the unicellular organism's knowledge of whether the molecule on its surface should be welcomed inside or stopped at the gates. All memories, even complex human ones, as well as our capacity to imagine possible states of the world, are extended versions of these very basic organismic capacities to *remember* that *too-hot* means that it is time to sweat and that *hungry* means that it is time to eat. In order to link these affectations with specific actions, memory is in constant relation to the present condition of the world. To even notice that the state of the environment has transitioned from cool, to warm, to hot, the organism needs to register a sequential or diachronic difference between these conditions. An organism that has the capacity to recognize the *tendency* (in this case, that the temperature is rising, that the thermal threshold is approaching) has to be structured in such a way as to *retain* this sequence in the order in which it has appeared. If it merely cognizes synchronic slices (now cold, now hot, now cool, now warm, now neutral), failing to grasp the sequence's ordering, it will fail to recognize the tendency at play, and may therefore fail to respond to an approaching threshold of sustainability.

In his *Confessions,* Augustine struggled with the apparent infeasibility of our capacity to order such sequences of events within the present. Though we presume that time is made up of past, present, and future, in what way, he asked, are the past and the future *real*? How is it that we can recognize that time has passed, that something has changed? Already Augustine was echoing Aristotle's idealization of time as being "in the soul," suggested in his conception of time as the number or measure of motion and change. Augustine similarly speculated that *only the present* is real, positing that the contents of the past and the future exist only in the mind, as memory and as prediction, respectively. Remembering and anticipating are aspects of the present. For God, all is absolutely present, all is *now,* for there could have been no moment *before* he created the world, as even time did not exist before this moment. But for we feeble mortals, some of that presence is obscured as pastness or futurity, and the best we can do to recover those other aspects of presence is to remember and to predict. Consciousness is, as it were, trapped in an arbitrary dimensionless sliver of the complex of presence. There is this sense in which time does not pass; it is still, eternal, immutable. And there is a sense in which time, in the soul's act of distention, is relative only to the mysterious workings of the mind.

Anyone who comes to speculate that we are confined to a mere snapshot, a cinematic-frame-like present, must explain how it is that we recognize changes and tendencies and how experience seems everywhere filled with intentional orientations. The present is never just a point. As Heidegger was to eventually argue, Aristotle's conception of time as a measure of change already concealed an idealization, an unwarranted privileging of the now as that which is spatially in our immediate presence, over that which has not yet come and that which has already receded into the past. For, moreover, it seems that even the present is always imbued with a content of duration. There are therefore two senses in which the present can be thought. As William James put it, there is a strict mathematical or objective present, which in each case resembles Augustine's durationless complex of time or Aristotle's numerical index of change, a single dimensionless point on the great complex of time, and then there is the experiential or *specious,* subjective present, a present that is always given as an integration of multiple events in sequence, Augustine's *distention of the soul* that forms a *tendency.* The organism's striving to remain within homeostatic bounds has everything to do with the second present, the polarized and oriented aspect of lived time: we need to retain enough pastness within the present to re-cognize change as a difference from the expected. The specious present is therefore *nonpunctate,* it is a *moment,* it has a *dénouement,* and it carries, as it were, its receding past and its incoming future within itself. James observed: "The lingerings of the past dropping successively away, and the incoming of the new, are the germs of memory and expectation, the retrospective and the prospective sense of time. They give that continuity to consciousness without which it could not be called a stream."[16] Similar reasoning on time was echoed by Franz Brentano, Husserl, and in his own way, Bergson. In other words, by the late nineteenth century, several thinkers were realizing the importance of considering how time as experienced differs from the snapshot or point. In order to make sense of internal, intentional, subjective experience, we have to posit time not as an iteration of discrete frames or points, but as *phases* of experience in momentary retention, each bleeding into the next. The question of how this *durée* of subjectively lived time is possible, or how it is that the present is retained long enough for consciousness to register the difference expressed by the incoming new, has since occupied many great minds.

Husserl's retentional model of time consciousness influentially described each moment of experience as split into three interacting aspects, the contents of which are continually evolving: the impression, the retention, and the protention. It is worth loosely following Husserl's retentional model for the next passages, not so as to insist on a phenomenological interpretation of the temporal, but rather in order to work toward a naturalization of the phenomenological perspective.[17] Husserl would of course have disagreed with this course of action, vehemently opposed as he was to naturalism. But we will see that Stiegler's extension of Husserlian temporal retention into the domain of *hypomnemata,* which implicitly already takes the first steps in such a naturalization of the *technotranscendental* conditions of the human experience of time, indeed puts into question the value of phenomenological or introspective perspectives on time, memory, and retention, suggesting that we are better off approaching the manifest and the scientific images holistically and in tandem. Eventually, it will become obvious that, if any further steps are taken in this direction, the question of the mnemonic construction of temporality is cracked open, as it were, and must be radically generalized for the nonhuman domain.

We cannot experience anything if we are not also constantly retaining some pastness with which to compare the given present, and the present never appears until it somehow disturbs or challenges our protention, the future-oriented anticipation of something yet to come. Every impression is in a constant state of passing into retention, but at the same time, the impression cannot even occur if it is not contrasted with that which is already retained. Husserl may have been too hasty in suggesting that the protention is a pure hospitality toward the future, for as Heidegger would later show, this kind of openness is strictly the exception. It is doubtful that our protentions are ever purely open and unbiased. Our protentions are rather driven and oriented: they derive, Jacob von Uexküll observed, from the organism's essential tropism toward homeostasis. As we continually synthesize temporal patterns into habits, it is in our constitution as organisms to expect our series of retentions to continue. It may indeed be impossible to retain a "pure" uncharacterized and undetermined retention of experiences, a pure openness to the future untainted by our expectations, for indeed, protention is essentially a skewed and biased relation to the future. And simultaneously, therefore, our protentions in turn condition what we retain. That which is retained is necessarily constrained by the protentions that skew our attentive openness. We do not remember everything that happened, nor do we register everything in our environment, only what our anticipations will have allowed to penetrate our retentional *Umwelt* and leave an impression on us. As organisms, our intrinsic homeostatic objectives condition our outlook, always preselecting what we expect to see, what we actually see, and what we finally will remember having seen.

Conscious experience or awareness is always found on the cusp of the present, the articulation between future and past, between this influx of novelty that we protend and the background of knowledge that we retain, but without being immediately aware in the present moment. The memorized and the forgotten meet at a point of indiscernibility on the *wave*

crest of experience, as James referred to it,[18] where knowledge and ignorance approach a limit from opposite directions. Samuel Butler astutely
evoked how this cusp of the present reflected a convergence of memory
and amnesia:

> It would, therefore, appear as though perfect knowledge and perfect igno-
> rance were extremes which meet and become indistinguishable from one
> another; so also perfect volition and perfect absence of volition, perfect
> memory and utter forgetfulness; for we are unconscious of knowing, willing,
> or remembering either from not yet having known or willed, or from knowing
> and willing so well and so intensely as to be no longer conscious of either.[19]

The remarkable discussions in Butler's *Life and Habit* suggest that, as
we progressively delegate what we learn to habit, what we learn ironically
retires from the active field of explicit, declarative experience and recedes
into the background of awareness, into what is today known as implicit
or procedural memory. The better we know something, the less we are
aware of it: it becomes invariant with regard to our conscious experience.
The better we become at a physically demanding task, the more we forget about it and take it for granted. There is this ironic sense in which
the very process of learning thus causes us to become ignorant of that
which we know best. And notice how this echoes precisely what Socrates
identified as the dangerous "pharmacological" status of writing, since
it co-opts our natural capacities for memorization and interiorization,
leading to their atrophy, a question extended to the whole of technics in
the work of Stiegler. It is that technologies, as extensions of our cognitive
strivings, carry with them the pharmacological inextricability of learning
and forgetting. There is a kind of forgetting that naturally accompanies
the influx of sensation into the retentional substrate: a *sampling* error. In
order to apprehend a new element of experience and integrate it into its
distributed network of retentions, the sentient system necessarily adopts
a new mnemonic topology, rerouting its circuits and adjusting its potential thresholds. Therefore, on the level of a single retentional context, the
modulation that carries the system from one state to the next involves an
immediate form of learning-forgetting that is identical to sensation itself.
And this is inseparable from what we characterize as the flow of time.

In the acquiring of procedural memory, learning is polarized: from
short-term memory to long-term memory, from difficult to easy, from
conscious and intentional to reflexive and automatic, or from explicit to
implicit. This first movement of cognition is indeed why the movement

of reason always takes the form of an *anamnesis*: it must recall what has been forgotten in the integrative flow; it must make explicit again what has been rendered implicit. Repetition here plays a fundamental role. It is repetition itself that envelops memory, that progressively *contracts events.* Practice makes perfect. In the central nervous system, the networks of neurons that cooperate in making a body move are progressively reinforced by repetition; they become increasingly integrated, interconnected, and interdependent. Gestural repetition causes the implicated presynaptic sensory neurons to be overstimulated, flooding them with serotonin, and when this accumulation reaches a given threshold, the sensory neuron not only transmits its signal to the motor neurons, but begins to manufacture new synaptic connections, creating shorter, denser, and more direct paths to the motor neurons. Hence the topology of the neural network changes: the brain reorganizes physiologically in favor of shorter circuits, making it more efficient at dissipating the *stress* of the overstimulating repetition. To the extent to which this process of learning through repetition is also a manner of implicitating conscious experience—that is, of rendering unconscious and autonomous the body's ongoing reactions to stimuli—it is also a kind of *forgetting,* the amnesia that necessarily conditions experience from behind the scenes. Deleuze characterizes this condition as *passive synthesis.* This first synthesis of time, in his view, happens automatically and implicitly, yet it occurs *asymmetrically,* ordering events in a series from future to past and from the general to the particular, constituting a polarized arrow of experienced time.[20] Each event is perhaps independent, but we passively contract these events into a polarized qualitiative impression.

Since life is temporally oriented, this contraction of events already implies a relative notion of past and future and gives sequential order to that which we will come to recognize as causality. The innumerable combined contractions that actualize us as organisms, by a continual unification and synthesis of the distinct elements that make up a repetitive series of events, force the many nested material "souls" that populate our bodies to *anticipate,* to *expect* the series to continue. In a word, they shape *habits.* We link disparate events into series and come to expect the series to continue, and we then delegate to background functions the task of contracting those series in order to open our contemplation to new events outside the series (though this happens passively, automatically, implicitly). Once the series is contracted as habit, any incoming elements that are already accounted for and already expected immediately slip into the background without registering a difference in conscious, explicit mentation. They merely *confirm* the series without rupturing our horizon of expectation. This constant habit formation corresponds to the mechanism that allows for causes to be linked to effects in Hume's analysis of causal induction. It also corresponds to the contemporary thermodynamic theory of cognition called "free energy reduction": we are composed of many nested *Markov blankets*, each a statistical model of the boundary that separates its process of expected-model maximization from the greater world.[21] The organism's repetitious production of habit in selectively observing conjunctions between events is the glue that binds causes and effects, and progressively constructs the organism's structure as a provisional model of its environment.

As a result, the passive synthesis is also the ground upon which the new declares itself: that which expresses itself as different from the last series, thereby initiating a new series of repetitions. The living present, nestled between past and future, corresponds to the organism's metastability between interior and exterior: its polarized membrane, its surface, *is* the cutting edge of its temporal experience. The learning mind is always to be found on that cusp between the outside, which is unknown, and the inside, which recedes into the background. Each event that makes its way into conscious awareness slices across the patterns previously repeated. It initiates a new convergence, a new world, populated by a new set of symmetries and invariances that immediately restructure and redefine the horizon of expectation by physically reorganizing the organism's boundary. Each world is defined by its divergence from other possible worlds, incompossible with the contingent facts that characterize the state of affairs after the event's updating of the series. Each event is incommensurable with the one it supersedes; yet it is supported by it, framed or contextualized by it.

The event, in other words, is *conditioned, not caused.* Rigorously speaking, there are no events in the *great outdoors,* just as there is no information in the wild: a mere point in spacetime is not yet an event unless it, like a bit of information, *makes a difference* for someone or something, for some model-maximization process; until then it is meaningless data. It has to express its difference from the prior series; it must rupture the topology of the various symmetries structuring the previous model of objectivity, which is equivalent to changing the self-organizing system itself. The organism's contraction of time, in its resolution of the differential tensions between the elements repeated within the series, creates

life
is
cognition

a kind of surface tension that supersedes the intercessions below, a web of reticulated *enabling constraints,* serving as the bootstraps that lift the agential process out of indifference. Thus it is only on this surface that new events now introduce their modifications. Simondon wrote that "the fact that a substance is in an exteriorized milieu signifies that this substance can befall, be proposed to assimilation, scar the living individual."[22] There is no event without attention or contemplation, no contemplation without these prior nested contractions, and no *facts* without the compossibility that they retroactively compose once integrated. The event is always a leap out of the causal regime that precedes it; it happens as a reordering, a restructuring of the network of causal lineages that constitutes the relative predictability and reliability of a world. Thus, in a sense, a new causal regime is created with every event, for the event is a catastrophe, a radical rewiring of the world model or model of objectivity, which is equivalent to the organism's expected difference from what it is not. This rerouting of the causal topology that co-constitutes agents and their world is a process inherent to all living things. This is why, as we saw in chapter 1, life *is* cognition,[23] and why Simondon can similarly claim:

All topological character has a chronological correlate, and inversely; therefore, for the living substance, the fact of being inside a selective polarized membrane signifies that this substance took hold in the condensed past.[24]

This "condensed past" of which Simondon speaks, which can be aligned with the "settled actual" in the work of Whitehead, constitutes a kind of *amnesia* that, in each case, must become the background of waking cognition, its blind face. If the "two levels" characteristic of human cognition—distinct and indistinct—are entangled, it is because the organism's reproductive structure is poised on a limit. Life is in perpetual transition, and it is this unification of stasis (perpetual) and process (transition) that, in effect, generates both memory and cognition on one side and a relative form of amnesia on the other. The asymptotic production of the organism's boundary between inside and outside, while never completed, is the work of a delicate balancing act. The structure of an organism *is* the retention of its past condition, a resistance in the face of a constant challenge by the thermodynamic forces that threaten its boundary, a continual attempt to minimize the difference between the world it expects and the world that actually expresses itself. Habit is indeed the essential property of life. While it never returns perfectly to the initial state, its failure to do so (the discrepancy between before and after) is incorporated as knowledge about the world: learning is trying, failing, adjusting, and trying anew. Whatever doesn't kill you makes you stronger. All this happens as an effect of primary retention, or what Brentano referred to as "original association" or "*proteraesthesis.*"

But, as I have already insinuated, retention is obtained at a cost. The organism must find the means to pursue this recalcitrance in the face of impending pressures, dividing the world in so doing into that which is of relevance for it and that which is not. Thus, in order to pursue its existence, in order to remain as it is for any period of time, it must simultaneously exclude part of the world for itself. The suppression and the retention are coimplicating aspects of a single process. In order to retain, we must exclude. The organism's constitution as an *extended critical state of matter,* both transient and stable, is in this way inseparable from the distinction between memory and amnesia. The living's perpetual phase transition implies bifurcations and nonlinearity, which indeed also imply an amnesia from the point of view of the superjective level: in order to focus on the task at hand, we must be able to *forget* that which is irrelevant to the task. Again, our contraction of sensory-motor phases produces an automated habit that no longer requires active attention. In habit, Hegel writes, "the soul has the content *in possession* . . . without sensation or consciousness" of the habit's content.[25] Heidegger's understanding of the concealment and deferral of technical being should be understood in a similar way. We must forget the hammer. We take a leap of faith and trust in the constitution of the causal framework that provisionally orders the scope of our actions. And yet the contents of this amnesia condition each task, in each moment, from behind the veil of concealment.

Thus, a kind of amnesia is inherent to the organism: it is implicit in the process that subtends its waking intentional agenda, that conditions perception and agency from behind the scenes. This is indeed the same implicit conditioning that subtends intelligibility and ensures the entanglement of the two levels of cognition, which is why, as we have seen, the intelligible model is never complete, why the syntactic offers incomplete coverage of the semantic. Again, intelligibility as such is necesserily correlated to the organism doing the synthesizing, producing habits, and integrating disparate events into its compossible world. Indeed, all these syntheses, these integrations of events *as* polarized and oriented, have a material condition.

If habituation and learning are analogous to a kind of *sedimentation* of repetitive series, gestures, and events into the provisional *ground* of the amnesia that conditions experience, it simultaneously clears a horizon of novel series to contract, new events to incorporate, new knowledge to learn. Hence, in learning, the content of the foreground is deposited into the background, ceding perpetually to new contractions of experience. This means that the ever-renewed surface is sensitive to its own distinct contingencies: it conditions that which can be manifested on the horizon, the ways the new event presents itself to actualization.

Secondary Integration

Thus, we obtain a second level. Deleuze referred to a *second synthesis* of time, and Husserl to a *secondary memory* or *reproduction*. Furthermore, we can enlist Whitehead's notion of *superject* in an analogous way: the secondary level of retention is the site of the emergent "soul" that pops out, as it were, from the combinations and contractions occurring below. It results from the resolution of reciprocal constraints between primary integrations, and is thus in some sense suspended on the surface of their contraction, identical to their mutual tension. (But of course, what I refer to as a soul is a material thing: *perceptronium* can be instantiated only in specific functional assemblages of matter, such as bodies, physiologies, deterministically constrained systems.) We see this in the expert pianist who, having practiced sufficiently to delegate her finger gestures to automated protentions of the musical sequence, is now poised to respond to more subtle variations in the interpretation of the musical work, such as varying emotional accents, looking for cues that are in excess of the series already accounted for. The contractions are still happening, but since the habit has been so firmly established and automated in procedural memory, it recedes into the background of awareness and her attention is free to consider other things. This freeing-up of attention is precisely how Husserl characterizes the *secondary* retention: "[It] is something free, a free running through: We can carry out the re-presentation 'more quickly' or 'more slowly,' more distinctly and explicitly or more confusedly, in a single lightning-like stroke or in articulated steps, and so on."[26]

The primary retentions and protentions now find themselves nested within a subsequent level of abstraction that allows for the comparison, in apparent real time, of different moments. This secondary structure encapsulates the first: it is not simply what allows time to be recognized as passing, as linked to the continuum or stream of consciousness, but what allows us to compare these phases in abstraction. In other words, the secondary retention contracts times that have *never been* present: representations, comparisons, correspondences, and various relations of resemblance and contiguity can occur only within this second synthesis of time, which, according to Deleuze, reveals the necessity of what Bergson called "pure past." In the Bergsonian interpretation, the present is the past in its most contracted state and the past is the general possibility of the present. Thus, the pure past is like the general context of all possible mnemonic events. Strictly speaking, therefore, habit and memory happen on two different levels. For Hegel, "habit is the mechanism of self-feeling, as memory is the mechanism of intelligence."[27] Somewhat analogously, for Deleuze, the first synthesis of time concerns the production of habit, while the second concerns memory proper. Perception and sensation are passive, at least from the point of view of the higher cognitive faculties: they happen automatically and without effort and constitute time as a passing present. The suborganisms that compose our multi-cellular organism, such as organs, cells, and organelles, may indeed be actively contracting their own events, but their contractions remain in the background of our waking consciousness as conditions for our awareness. Bergson's paradox of the present (that it constitutes time while also passing within it) implies that "there must be another time in which the first synthesis of time can occur."[28] Habit, in other words, "is the originary synthesis of time, which constitutes the life of the passing present," while memory "constitutes the being of the past (that which causes the present to pass)."[29] There is thus another synthesis that, at least provisionally, "grounds" time, and this ground is memory proper. If retention itself can be said to occur on the level of a singular substrate of inscription, there is a sense in which memory is a process concerning at least two, but usually many, different retentional substrates in interaction and cooperation. Memory is a complex of retentions. And if memory grounds time, it is because the secondary retention and protention have a selective effect on the primary level. There is a retroactive effect, a top-down influence, where our secondary retentions, which allow a free movement between the different elements of the primary level, condition or select which primary retentions will happen to be sampled: our secondary retentions constrain our primary retentions and, in this sense, must be said to logically precede the primary retentions.

We might say that there are, because of this, no truly "empty intentions"; they are always already infected with prior selections. This is the feature of the ontology of lived time that inevitably complicates the idea of the *given*, for it seems we can never really say that we have truly grasped the given as such, the true and original sense datum that motivates a thought, as it is always already tainted with prior conditioning. Thus, in a sense, not only is the noumenon hidden away by the phenomenon, but even the phenomenon itself is always already obscured and transformed before we grasp it consciously or explicitly, obscured and transformed by the layers of top-down *mereological* influence that characterizes the nature of temporal perception.

When the experienced pianist delegates a coordinated series of complex finger gestures to habit, she has to "automate," and thus in this sense "forget," the specific positions of her fingers on the keyboard. The amateur, by contrast, does not succeed in disconnecting the task of articulating his fingers on the keys, and if he succeeds at all in playing the melody, his interpretation sounds flat and mechanical. This is because the inexperienced pianist fails to liberate the second synthesis by outsourcing his gestures to the habits of the first. The practice and repetition of playing the piano eventually produces a habit, which is a short-circuiting of discursive, declarative, explicit experience: the high-level conscious retentions are pushed past the boundary of the phenomenal, beyond the infraceptive limit, into the infinitesimal time scales that escape our experiential present. The kind of forgetting taking place here concerns not only the modulation of a single retentional substrate, but an *interaction* between two levels. When occupying the more general mnemonic context of the second synthesis, the habits, like closed algorithmic procedures, will populate the peripheries of a hybrid differential retention (the mnemonic integration). The automated finger gestures are re-presented, compressed in the ongoing flux of the secondary context. A consciously demanding task requiring the coordination of many sensory-motor subsystems has been delegated to a shorter circuit on the periphery of explicit discursive experience.

A habit, a *procedural* memory, has been formed as the product of repetition. Through the repetition of sequences of postures and gestures, the mnemotechnical assemblage operates a *supersession* of the repeated series, a contraction of the series into an infinitessimal time, shorter than the shortest lapse of time thinkable within the discursive perceptive level, pushing it beyond the periphery of direct awareness. And yet our current state of awareness is supported by this constant passive learning-forgetting, which conditions our immediate experience of the past and future as withdrawn from the present. Brian Massumi suggests that human experience is suspended across this process of oscillation between retention and release, which is why it appears to us as recessive or passive.[30] This echoes Deleuze's characterization of his notion of "virtual" as occupying a time circuit both shorter than a single experience's frame of reference and longer than a single thought's capacity to retain.[31] Yet this implicit or virtual tension subsists within every waking experience. It has a functional role in the constitution of the ongoing metastable and polarized criticality of life and cognition.

Thus, the topology that defines the memory of a living being as a set of *satisficing* responses to given environmental stimuli is the *material* condition of its experience of time. Simondon writes, "the schemes of chronology and topology can be applied to each other; they are not distinct, and form the primary dimension of the living."[32] Living beings always implement some form of reason or intelligence: they *conditionalize their inferences* about the world on the *state of affairs,* on the events that present themselves to their sensory apparatus. Simondon understands that subjective temporal experience mirrors the organism's necessarily polarized structure.

> On the level of the polarized membrane the interior past and the future exterior confront each other: this confrontation in the operation of selective assimilation is the present of the living, that is made of this polarity of passage and refusal, between substances past and substances that are to come, presents one to the other through the operation of individuation; the present is this metastability of the rapport between inside and outside, past and future.[33]

But the second integration complicates this picture. We humans are more than just a single membrane: we are composed of many tissues and membranes, many retentional substrates, many Markov blankets and world models. Consider how humans have the peculiar ability to displace conscious attention from sensation to reflection. This testifies to our remarkably flexible metastability as complex organisms. If the causal world is much like a field of deterministically linked action–reaction events, each successive layer of complexity that life has evolved implies the loosening of such links and bonds, where each action–reaction event is interceded and mediated by many mnemonic buffers, delays, and retentions. If the first organism had already gained more causal "wiggle room" than its nonliving environmental surroundings, the more complex organisms have simply continued this trend: the human's metastabilization across a bio-psycho-linguo-technical spectrum of retentional contexts affords it that much more room to wiggle, that much more flexibility and freedom in relation to its causal constraints, due to various action–reaction *buffers* that effectively allow for "decisions" pertaining to an array of "choices" about how to respond to given stimuli. This corresponds to the well-known narrative according to which life is *negentropic,* or more accurately *anti-entropic* (as Guiseppe Longo and Maël Montévil remind us[34]): it locally resists and delays the incessant movement of the cosmos toward disintegration and entropy.

All this indeed manifests itself in the characteristic *freedom* of the secondary integration, the capacity of human attention to be transposed toward superior or inferior strata (macro or micro perspectives) of a given time series. The more we "contract" a repetition, the more we come to expect the next iteration, and hence the more we are surprised by an interruption in the sequence. Bayesian interpretations of cognition will hence suggest that, as we contract events and repetitions, we are simultaneously optimizing our predictive model: the habitual fades into the background of experience because our probabilistic predictive model accounts for the events in the series as repetitions of known or expected characteristics. The first time we hear a given melody, each new subsequent note is completely unknown; but, when the melody repeats, the repetition contracts and we are progressively habituated and come to expect its sequence. We no longer hear it as a series of distinct notes, but as a unified musical motif. The contraction of the series of notes into a refrain liberates, in that instant, a new modality of hearing: our attention is free to transpose itself toward another stratum of the composition, one where we no longer *sensuously* hear the notes and their relative intervals, but are attentive rather to the melody, the molar entity they compose. We are now contracting variation on a secondary level, on the level of the differential tension between the verse and the refrain, rather than between individual notes. The object of hearing on the level of the series of notes becomes the subject of hearing on the level of the interposed melodies. And there are also examples of the reverse transposition of awareness, the contraction toward an inferior horizon: for instance, the scholar reading a difficult text for the first time who retains only the coarse-grain meaning on an initial reading and then remarks the subtle nuances of the argument on a second, more thorough pass. The molar traits are thus contracted as attention is displaced toward the inferior limit, passing into the molecular, where one can contract the *sub-text.* Echoing the myth of Actaeon, the predator becomes the prey, the hunter becomes the hunted.[35]

As the series of operations is contracted into habit on the first level, as synaptic connections are formed to reduce reaction times, and as our elementary cognitive subsystems correct discrepancies between their predictive models and the incoming sensory data, the organism is swept up in a process of exteriorization and implicitation that mirrors the characteristics we usually reserve for mnemotechnical considerations, long before technology is even involved. As short-term memory transitions to long-term memory, as conscious mentation is transformed into procedural automatisms, our horizons of expectation are correspondingly modulated. Indeed, the more we implicitate the explicit, the more we short-circuit surprise by fine-tuning our predictive models to account for the unexpected, the more in the world we take for granted.

Nowhere is this complex interplay between levels of retention demonstrated more vividly than in the tragic case of Clive Wearing. After he contracted a severe case of herpes encephalitis that destroyed his hippocampus, the accomplished British musician, composer, conductor, tenor, and pianist—a highly erudite, intelligent, and witty man—became profoundly amnesic, both anterograde and retrograde. Having lost his former episodic and (much of his) semantic memory, as well as his capacity to form *new* episodic and semantic memories, he became trapped in a tiny sliver of the present, a variable window somewhere between seven and thirty seconds, continually feeling as though he were just waking up for the first time, incessantly being surprised to find himself where he was. Indeed, the ability to be surprised is inversely related to memory and expectation: the more one remembers and expects, the less one is surprised, as when the child's astonishment and awe wears thin in the accumulation of successive similar experiences. The anterograde amnesiac's conscious experience is reduced to a disorganized flow of primary retentions, and he becomes incapable of integrating multiple moments into the conscious whole of a secondary integration. Wearing's diary is consequently filled with momentary episodes of waking up for the first time, as if from a coma: "now I am awake," "now I am really awake," "now I am truly alive," "now I am conscious for the first time," and so on. He once stared at a piece of chocolate, repeatedly covering and uncovering it with his hand, caught in the loop of seeing this same mundane item with a perpetually renewed sense of wonder: "Look! . . . It's new! . . . No . . . look! It's changed. It wasn't like that before . . . Look! It's different again! How do they do it?"[36]

And yet, throughout this, Wearing retained his musical ability, still singing, conducting a choir, sight-reading, and playing the piano with grace and passion. His facility with the keyboard, like his agile and witty use of language, had been hardwired into his brain through the force of repetition, transferred from declarative to procedural circuits, and these were among the parts of his brain that survived the encephalitis. The retentions of his conscious high-level experience, his constantly renewed sense of awakening, were no longer coordinated with the lower-level retentions, the micro perceptions-actuations articulating his fingers on the keys. The retentions above and below no longer integrated into the necessarily multiplicious synthesis of memory proper. "I have no thoughts," "I am not conscious," he claims in some of his darker moments. He thus forever strives for continuity, perched as he is, on "a tiny platform . . . above the abyss,"[37] his only chance being to continue talking, loosely linking one idea to another, creating chains of puns that keep him in the flow from one moment to the next, keeping his little platform afloat.

As complex multilevel life forms, it is in our very nature to "automate" certain reactions to sensation. The brain's lower cortical strata, which maintain vital bodily functions, operate quite well without explicit guidance by our conscious explicit mentation. Only when events in the environment push these systems out of their zones of comfort must the lower levels bother the higher ones, transmitting action potentials, producing conscious reactions like pain, hunger, fear, and excitement. This perpetual adjustment between expectation and world, this conditionalization of inferences on facts retrieved through trial and error, is the prototechnical, biocognitive analog of the transition between Heidegger's *presence-at-hand* and *readiness-to-hand*. When our experience is implicitated in automation, when it recedes into the background, the individual molecular operations constructing our waking moment are sublated, assimilated, and statistically unified as an autonomous molar entity: in the mnemonic event, the *little perceptions* of experience are converted from presence to readiness; they have acquired a tool-like being for the "purposes" of framing and orienting of the organic present. So, even after Wearing lost the ability to create long-term episodic or semantic memories, all the physiologically registered automated capacities he had long trained for remained in their full *readiness*; he was just no longer in conscious control or possession of them. He plays the keys and sings the melody with passion, but from moment to moment a constantly reiterated Wearing appears into those hands, into that voice, with no possibility of continuity. If someone directs him, for instance, to practice a new piece of music repeatedly, he is able to contract new *procedural memories,* even though he has no recollection of ever having played or practiced the piece. And when the piece of music finishes, he is overwhelmed with a fit of convulsions as the lower retentional strata, now *ready* for further instruction, try to signal the higher ones through neural circuits that no longer exist, resulting in chaotic physical and emotional impulses.

Thus we recognize the same distinction between lower and higher that has echoed throughout our earlier investigation: the higher level of explicit and distinct knowledge; the lower level of indistinct and confused cognitions; the higher level of technological or survival related concerns; the lower level of aesthetic concerns. For, it becomes possible and compelling to speculate from these observations that the perceived differences in the human domain between these levels has much to do with the *mereological* structure of organisms, which are composed of innumerable nested *holons* that instantiate their own *little perceptions,* and which then become fused, con-fused, in their signaling to and coordination in the higher level, the level of the relative whole the parts collectively compose. One of the defining aspects of the organism is precisely that there are top-down influences and control mechanisms in which the individuals of one stratum submit to the "decisions" of the molar entity they collectively form one level up the hierarchy. This molar entity does not "make its decisions" based on the individual desires of each holon in the set, but based on some form of statistical interpretation, a macroscopic, coarse-grained interface with the world. This coarse-grainedness is to be thought of as the higher level's own relative amnesia: it does not see the real, but is constrained to see only what the transcendental structures of its relative subjectivity (the structurally coupled holons it is composed of) collectively allow it to see. To return to a helpful example given in chapter 1, we cannot explicitly specify the individual marks between two *just-noticeably different* shades of color because the system in our organism responsible for this report is constitutively coarse-grained, synthesized from all those little perceptions. Like the spokes on Leibniz's cogwheel, the molecular perceptions must be con-fused into the molar entity that constitutes our discursive report of the color difference. Specifying the individual molecular perceptions therefore remains constitutively "beyond" this molar system's capacity: the molecular perceptions are its immediate transcendental structures conditioning its own subjective embedding and orientation within the environment.

Tertiary Integration

We have so far discussed the two levels of mnemonic retention/protention that immediately express themselves in the analysis of human temporal experience. A first level structures time and causality as a one-way street and synthesizes repetitive events into habits. A second level allows for comparisons and contrasts between moments of the first, corresponding to recollection and imagination. There is a correspondence between the temporal characteristics of the relation between these two interacting retentional levels and the indistinct–distinct dichotomy that has served as our point of entry into the questions of aesthesis, noesis, and prosthesis. As I have argued, these two levels are instantiated in the organism as functional capacities inscribed in the materiality of the extended critical states of life, which is essentially in the business of *reasoning*, of adjusting its model of objectivity to the continual flow of facts it registers. I have suggested that both levels, though they allow us to cognize elements of the real, are also founded on relative amnesias, blind-spots, takings-for-granted, and over-lookings that, in an abstract way, mirror Heidegger's understanding of *tool being* as a pragmatic dichotomy between the explicit and implicit.

As Heinz von Foerster remarked, once we have a second order (system or cybernetics), there is no need to go any further: third, fourth, fifth, or Nth levels of order are immediately presupposed by the possibilities opened up by the reflection of the first in the second and the second in the first, because one has "stepped into the circle that closes upon itself."[38] But Stiegler has nevertheless (and rightly, one might argue) found the need to posit such a third order for describing the specifically human predicament. For indeed, with regard to the human context, where do actual tools, actual human technologies, media environments, practices, and artifacts fit into the retentional picture we have been discussing? Stiegler's great contribution to the question of memory and amnesia enters the scene. In addition to Husserl's two levels of retention/protention, Stiegler's *Technics and Time* introduced a tertiary level corresponding the human's investment of technologically exteriorized mnemomnic supports, or *hypomnemata*. Though Husserl had not thought of technologies as constituting a third level of temporal synthesis and conditioning, Stiegler, after Derrida, interprets Husserl's *Origin of Geometry* as having shown that history begins "as the history of geometry, which is also that of an instrumentality: geometry [and, therefore, ideality] is not conceivable outside of a process of communitization made possible by a *technics of presentation* of the 'already there': no geometry without instrumental retentionality; without constitutive tertiary memory."[39] Thus, in another sense, the third level of retention/protention actually comes first, as it recursively conditions the first and second levels. The individual states that a mind can take on (the thoughts, experiences, and cogitations that can compose our interior life) are directly and operationally interlocked with the deterministic constraints of our technological environments. The kinds of subjectivities we inhabit are directly influenced by the *techno-transcendental* conditions of thought.

This corresponds to our concerns in chapter 1 regarding the contingent *construction* of knowledge. What Stiegler stresses is that this construction necessarily depends on the exteriorized memory support, the *hypomnematon*. His foremost contribution has been to insist that technics condition our most intimate experience, and even the seemingly a priori conditions of time and space. In this, he has radically extended what Derrida already controversially claimed about writing—namely, that it *logically* precedes the spoken word and the oral tradition—to the whole of technics, which compelled Stiegler to commit his philosophical project to "general organology," an account of how various forms of technics promote or atrophy the conditions, priorities, and possibilities of the psychosocial.

Humans are largely defined by the social externalization of the individual's memory into shared artificial devices and practices. According to André Leroi-Gourhan, it is the progressive prosthesis of the hand through tools and the mouth through language that led to the mode of survival particular to the human species, constituting both the social realm and the subjective, psychological realm. This notion that technological environments condition our most intimate temporal experiences has now become one of the primary tenets of contemporary philosophy of technology. The thesis of *originary technicity*, with its roots in the work of Heidegger, Leroi-Gourhan, and Derrida and greatly developed by Stiegler, implies that we should not understand technologies as mere instruments in the service of human wills and desires, but rather as part of the conditioning principles that constitute us as human beings, along with our social *habitus*, our physiological structures, and the modes of cognition they enable. We know, for instance, that our early ancestors lived in a technological environment that was relatively stable, changing very little over two million years from the Oldowan to the Aurignacian, during which several distinct species of hominid evolved and went extinct without incurring any substantial change to the industrial production of stone tools. In a sense, it is not we who have created our tools, but our tools that have created us: since the beginning, humans have evolved with and through technologies. The thesis of originary technicity hence suggests that technologies and technical environments continue to participate at every step in our constitution as cognitive and social agents, including our intimate temporal experience (our horizons of expectation and attitudes toward the future).

To understand how the third level of retention/protention constituted by technology's *organization of inorganic matter* situates itself within the decidedly *organic* picture of the two previous levels of retention, it is useful to take a short detour through Plato, to whom Stiegler continually returns. Recall that, in the *Meno,* Socrates insists he knows nothing about virtue, but also claims that, through dialog and interrogation, by asking the right questions *in the right way,* the nature of virtue can reveal itself. To prove his point, he calls on an uneducated slave boy from Meno's entourage, asking him whether he knows how to find the double of the area of a square. By drawing in the sand, first a square, then diagonal lines leading from the square's corners, Socrates demonstrates that the slave boy "spontaneously" recovers the solution to the problem. In other words, it is by *exteriorizing* the problem as a *hypomnesic* inscription within the inorganic domain (the sand) that the boy can learn what he previously did not know: *new knowledge* seems to spring forth from the exteriorization itself. In an apparent blurring of the analytic–synthetic distinction, without the exteriorization and its subsequent synthesis with the concept of the square's area, it would not have been obvious that the truth of the area of the doubled square was in fact already contained in the proposition itself: the analytic definition of the square reveals itself only after the retroactive synthesis of the exteriorization and the concept.

Socrates proposes a mechanism for just how this externalized memory is supposed to have allowed the slave to discover the answer. Meno first challenges Socrates with a paradox: how do you know you have really found what you were looking for? How is recognizing new knowledge as new really possible if you did not know it already? And if you *did* know it already, in what sense did you really learn anything? The famous response offered by Socrates concerns a method of communication between his realm of ideal types and the earthly, mortal realm of tokens and particulars: the doctrine of *anamnesis.* The Socratic concept of anamnesis, or reminiscence, is articulated on a metaphysics of reincarnation. "Seeing then that the soul is immortal and has been born many times, and has beheld all things both in this world and in the nether realms, she has acquired knowledge of all and everything; so that it is no wonder that she should be able to recollect all that she knew before about virtue and other things . . . since, it would seem, research and learning are wholly recollection."[40] With each successive reincarnation, the soul forgets the universal truths it has access to. And so, for Socrates, the experience of novelty is always a recollection of that which our souls have forgotten in the amnesia induced by rebirth.

Now, it is obvious that this distribution of mnesis, amnesis, and anamnesis correlates to the dialectics of learning and forgetting we have been touching on thus far. But, of course, the Socratic understanding of anamnesis is dependent on his metaphysics of reincarnation, rather than on materialist or naturalizing account agreeable with the one we have been pursuing. However, in the *Phaedrus,* another dialogue from which Stiegler draws inspiration, we do find the rudiments of a materialist conception of technical amnesia and anamnesia. The dialog contains another important discussion about memory, but this time about memory's relation to writing, another exteriorized memory support. The argument pits writing against the oral tradition in terms of what they each allow to record and recollect. Again, it is here that Socrates famously compares writing to a *pharmakon,* which implied, as Derrida showed, that writing is both good and bad, both medicine and poison, for though it may allow one to remember more words through mnemonic exteriorization, it nevertheless atrophies one's innate capacity to internalize those words.

Everyone is familiar with this kind of technologically induced amnesia. With the adoption of the technology of the chair, for instance, Western culture has forgotten how to squat. This is because memory and knowledge are functionally instantiated within organisms: they are processes inseparable from *practices,* from the series of operations required for constructing knowledge. To know is to do, and without the doing, the knowledge disappears; to remember is to have a certain capacity to adopt a physiological state. The "memory" of how to squat is physiologically inscribed in the flexibility of the Achilles tendon, and in the range of motion of the ankle; the fact that most in contemporary Western societies lack that flexibility, probably as a result of the physiological supplementation provided by the chair, merely demonstrates that memory is identical to the organism's capacity to take on specific system states in response to environmental stimulations (this was magnificently demonstrated in Eric Kandel's work with marine snails).[41] We are all ensnared by this technical redistribution of memory and amnesia. There is no escape. Two decades ago, for instance, everyone in the Western world memorized the phone numbers of their immediate friends and family, but as phones have gotten "smarter," we have become reliant on the exteriorized memory support these devices offer, a contemporary echo of the same technical redistribution of mnemonic capacitates that Socrates discusses in the *Phaedrus*. Here, instead of outsourcing demanding tasks to the automated operations of our inferior retentional levels as habits, we outsource them to our external technological milieu.

Now, it is telling that, in the two dialogues, the *Meno* and the *Phaedrus*, Socrates holds what appear to be two contradictory positions: in the *Meno* he demonstrates that exteriorizing is good, as it allows the slave boy to discover something he previously did not know, by simply making geometric inscriptions in the sand; in the *Phaedrus*, he argues against exteriorization as a poison that will eventually render humans less capable of thinking for themselves. This is probably due to Plato (or Socrates) having not conceived of writing and geometry as conceptually allied forms of mnemotechnics. But ultimately, the point that has arisen out of this philosophical tradition is that geometry, writing, and indeed all technics and media are marked by the pharmacological character of their mnemonic redistribution of subjectivities across retentional substrates. They are remedies in their solution to technical problems, yet also *poisons* inducing various atrophies of prior capacities. This is mirrored in the notion of *prosthesis* itself: the technical prosthesis is not merely an extension of the body (as emphasized by Marshall McLuhan); it also implies the *replacement of a lost limb*. It is a *supplement* in the Derridean sense: both an enhancement and a substitute. Thus, in their mnemonic dimension, what characterizes technics is how they allow us to memorize or learn while simultaneously causing amnesia and ignorance.

We can therefore begin to naturalize the Socratic metaphysics of reincarnation by simply invoking the materiality of prosthetic supplementation, by replacing the amnesia induced by rebirth with the amnesia inflicted by successive evolutionary and technical stages, each involving an irreducible redistribution of inclusion and exclusion, learning and forgetting, and distinct and indistinct knowledge. So, again, primary retention or contraction itself can be said to occur within a single retentional context, as an inscription into the critical system, poised as it is on the edge of a looming incursion of difference. Memory, however, can perhaps be distinguished from this primary retention in that it inevitably concerns the integration and reciprocal dependence of multiple such retentional contexts. This distinction corresponds to what we saw a bit earlier: Deleuze's first synthesis of habit is not to be confused with his second synthesis of memory. Memory is a step up from the simple contraction of events happening in each retentional context: it concerns their comparison and their contrast, and eventually their coordination. The tertiary retention provided by externalized memory supports in Stiegler's paradigm retroactively conditions the activities of the first two levels. In fact, since each retention is doubled by a correlated protention, the levels condition each other, infecting each other with biases. This leads to a *permeability*[42] between the three levels, making it difficult to disentangle them to any faithful degree, and this task indeed underscores one of the central challenges of Stiegler's general organology: how can we faithfully come to understand what are effectively the technotranscendental conditions of our very capacities of understanding? For, though tertiary retention is third in relation to the other two, in some sense it actually comes *first*. Just as writing is logically prior to speech in Derrida, so is mnemonic exteriorization the necessary precondition of internal time consciousness in Stiegler. "Human memory is originally exteriorized, which means it is technical from the start."[43] There is no interior, no internal experience or consciousness, before the outside is first invested, discretized, and inscribed, before the external retentional substrate operates its retroactive decoding of the interior. It is through this exteriorization that the subjective interior progressively emerges in compensation. The other two levels of human retention can be said, in this retroactive sense, to be constituted as responses to the original acts of drawing inscriptions in the sand and of scoring tally marks in bones and stones.

Protomnemotechnics

Stiegler has influentially discussed the nature of technics as a "third layer of memory," externally supervening on the primary and secondary forms of retention that go on within the human organism. All technics, not just mnemotechnics proper (technical practices consciously used to store memories), are bound up in the complex weave of retentions and protentions that constitute human recollections, desires, anticipations, identities, priorities, and so on. He has identified this logical priority of mnemonic exteriorization as the paradoxical *originless* origin of homi-nization, the human's "default of origin." As in Derrida's account of the *arche-trace*, whose very structure is self-effacement, in Stiegler's account, "one memorizes only by forgetting,"[44] only by exteriorizing memory and investing the inorganic with *hypomnesis*, implying that humans are con-stituted by an "originary technicity." All the traces we leave outside per-ception, outside the body, all our exteriorizations into the artificial realm, written or otherwise, retroactively participate in our most intimate con-stitution, shaping our drives and desires.

Stiegler's theory notably centers on a paleontological reading of Derrida's arche-trace, which he derives from Leroi-Gourhan. He hence understands the emergence of technical exteriorization as the event that will have led to hominization as a bifurcation from natural processes of evolution. And thus, he entertains a fairly strict distinction between human life, which invests technological memory supports and pursues biological evolution through artificial inorganic means (technical adop-tion), and the prehuman animal or vegetable life, which does not exterior-ize itself in such a way. Humans are born into environments and cultures that condition the expression of their predispositions. Each new techno-logical innovation merely continues this history of hominization as a per-petual exteriorization. The human individual externalizes its physiological memory into an evolving topology of distributed, multilevel mnemonic constraints, allowing the psychosocial system to store more memory by affording it new degrees of variational freedom. But is this really so excep-tional in nature?

We humans, it is true, have constructed a vast and complex artifi-cial milieu to inhabit, but it is no more artificial than the cathedral-like mounds that termites build. The anthill and the termite mound are oft-cited nonhuman systems whose organizational structure also relies on an explicitly externalized memory, information that is stored elsewhere than in the individual agents of the species. The features and topology of the termite mound are *hypomnemata* for the termite, as they implement an externalized and intergenerational mnemotechnics—the accumulation of a field of pheromones varying in intensity from point to point—which in turn instructs the termites about how to deploy themselves individually. Even in creatures that do not possess a nervous system, such as the slime mold (*Physarum polycephalum*), an amorphous blob-like superorganism made up of individual unicellular creatures, there is evidence of exter-nalized memory. The brainless slime mold's behavior, according to one study, "strongly suggests that it can sense extracellular slime upon contact, and uses its presence as an *externalized spatial memory* system to recognize and avoid areas it has already explored." Because of this externalized memory, the slime mold holistically exhibits intelligent behavior: it solves navigation problems that remain challenging in robotics, finds its way out of labyrinths, and successfully optimizes shortest-path problems. Since it does all this while being so simple in relation to "higher" complex organisms that are composed of multiple interacting systems with specialized cells and tissue structures, such as those making up the central nervous system, the study suggests that "*externalized spatial memory* may be the functional precursor to the internal memory of higher organisms."[45] This ironically lends *nonhuman* support to Stiegler's claim that exteriorization "comes first" in relation to the primary and secondary levels of retention. So, if there are such examples of mnemonic exteriorization in nonhuman nature, then are humans truly that exceptional?

Stiegler himself has tackled this question. He admits: "In the case of ants, pheromones are the chemical traces inscribed on the ants' habitat as support—the anthill and the surrounding pathways marked by individual hunters—and as a mapping of the collective. Exteriorized memory is already clearly evident here: an inorganic support and the resulting organization of territory (the anthill itself)." But crucially, he argues, the difference between this kind of externalized memory and that of the human is that, in the human, it participates in a "transindividuality" between individuals. In the case of ants, he argues, each ant has no individual memory, and so what happens on the level of the organism cannot be said to be transindividual. So, he reasons, in the case of the anthill, we cannot speak of *epiphylogenesis*: "There is no question of a transmission of individual or collective experience. And this is also why there is no organization of the inorganic through a technical tendency—but rather a structural coupling of the animal group and its surroundings."[46] It is as though the ant is *completely* exteriorized into its hypomnematon, the anthill and colony; in other words, its exteriorization does not fold back upon itself to furnish the individual with an interiority or subjectivity.

This observation indirectly reveals an interesting prospect: that the human predicament corresponds to a certain way of being *metastabilized across several retentional strata.* Human aesthesia indeed seems to be a question of operational coordination between multiple retentional substrates. But, rather than illuminating the human condition specifically, does this metastable distribution across several nested retentional contexts not more generally also describe the organism's fractal structure of mereologically nested systems? Complex organisms are made up of many interacting systems in which influences travel not only up the hierarchy but also downward from the higher levels (say from the brain's higher cortical strata to the efferent neurons): in this way the whole constrains the expression of the parts. There is thus a kind of resonance, a *mereological bio-resonance*, among the retentional substrates that characterize the integrated aesthesia of the complex organism. It is the trans-individuating reticular network emerging from the nexus of interactions among many critical systems, each oriented in space and time, each being its own individuating process. All organisms are distributed differently across these retentional substrates.

We find here a certain continuity between biological prosthesis and artificial prosthesis that corresponds to the transductive processes Simondon found were shared by self-organizing individuations of different scales and complexities, from crystals to biology, onward to the techno-psycho-social system. This observation has the effect of modifying the exceptionalist account Stiegler gives of the origin of technology. From a systems-analysis point of view, we might understand the progressive hominization of our evolutionary phylum, characterized by the externalization of memory into artificial objects and practices, as the emergence of a supplementary level of complexity founded on and pursuing the biological's impetus toward exteriorization. All individuations work within a mnemonic economy consisting of information exchanges between critical systems and their environment. There is thus a long chain of evolutionary forms of *prototechnical* adaptations and metastabilizations that each emerge through a double movement of reticulation and exteriorization, intercession and supersession, aesthesis and prosthesis. The real value in Stiegler's insight into the technotranscendental constitution of the internal experience of time is thus not its establishing a strict human exceptionalism, but its demonstrating the contingent external factors that lead to the metastabilization of cooperating retentional substrates, from region to region, from organism to organism.

Take a very simple self-organized critical system: the pile of sand. As was discovered by Per Bak and his colleagues, a pile of sand or rice is a surprisingly interesting phenomenon.[47] It is perpetually poised on a looming catastrophic event—the avalanche—and yet it perpetually self-organizes such that it maintains its characteristic structure. One single additional grain of sand will cause the pile to readjust, sending cascades of grain down its surface. But, after the avalanche, it will have regained its meta-stability on this evental cusp, and it will have self-organized on this critical edge. The pile of sand can therefore be said to satisfy the criteria for primary retention/protention. Its system structure is that of a *poisedness* between the past and the future: it is oriented in time, for it negentropically pushes back against the pressures of the second law, which wants only its disintegration. So, though it contracts events into its holistic systemic structure and, in a minimal sense, thus exhibits a form of *habit,* it does not satisfy the criteria of secondary retention, or *memory.* This is the difference between such a structure and what Francis Bailly and Longo call *extended* criticality, which characterizes biological systems. The pile of sand cannot differentiate between one event and another: the future is always avalanche and the past is always resistance to avalanche. It is always selecting the same singular aspect of its environment, no matter what comes its way. Though it does select, retain, and protend, it cannot change its relation to the environment: it cannot modulate what it selects, retains, and protends. This is why Deleuze says that a second synthesis of time is required for the reproduction, representation, and association (according to relations of resemblance and contiguity) of multiple pasts.[48] The secondary temporal contraction allows the critical system to compare one event to another, thus permitting a landscape of possibilities to open up before it. The leap into secondary retention thus corresponds to an exponential increase in the system's degrees of variational freedom: the second synthesis or secondary retention is a correlate of its expanded phase space. It is the set of abstractions required for navigating and engaging in a *world* that is more than an implacable one-way flow of time. Time is now cracked open, fanning out from the present as an array of possibilities and their infinite ramifications.

The passage from a self-organized critical system to a self-organized extended critical system corresponds, I contend, to a transition from the first to the second synthesis, and this transition itself requires a *retentional exteriorization,* not unlike that which describes the passage from secondary to tertiary retention in Stiegler's account. We observe such a pattern in the living cell. The various organelles that compose the cell (mitochondrea, vacuoles, lysosomes), each its own functionally integrated organic machine, come together in symbiotic cooperation, forming a common boundary (the cell membrane) that regulates and limits their interactions and collectively defends them from entropic forces. Each one's retentions are hence *externalized* in this sense, toward the *super organism* that the cell already, in some sense, is. None of the organelles can survive outside of the cell: they are now codependent and reciprocally presupposed, reliant on the mnemo-bio-technical nexus of integrated cooperation characterizing the cell as a whole. Its emergence marks a transition: a new disjunctive unity of the whole has supplemented the plurality of the incommensurable causal recursions that define each individual organelle as a relative primary-retentional context. In the "symbiotic assemblage," a molar *supersession* has compensated the molecular *intercession*, reflecting what Luciana Parisi describes as "a bifurcation of matter into two parallel machines of organization."[49] The reticulating *aesthesis* has been redoubled by a supplementing *prosthesis*. On its own, each part is a retentional context of the first degree, a contraction of events as habit, poised on the cusp between past and future; but it is only on the level of the exteriorized whole that the secondary retention takes hold. With every emergent level of organization, this *redistribution* of roles takes place: the product of endosymbiotic participations among ever more complex causal regimes externalizes its process, automates its functions and entropic delay mechanisms by cooperating with other causal circuits, and thus loses part of itself, while also gaining greater efficiency (freeing itself of some of the burden by sharing it with others).

The living being, therefore, might be said to be composed of *at least* two, but often many, levels of integrated structure and functionality, a minimal hierarchy of complex interdependencies among retentional substrates. The pile of sand is indeed a critical, selective, asymmetrically poised, and antientropically organized structure: it resists the arrow of time in a minimal sense. We have to take this seriously, for an inorganic self-organized critical system may represent the limit case of aesthesis. Memory, however, specifically concerns a coordinated behavior among retentional contexts. This coordinated behavior refers to the mereological topology of the organism, which constitutes it as an *inference* machine. Memory is allied with expectation, just as retention is inseparable from protention. Since selections and inferences are articulated on some kind of memory (some sort of stored information that needs to be compared with contingent events), and since this memory must be structurally embedded (actualized) within the organizational degrees of freedom of the system doing the selecting, such a "cybernetic" perspective illuminates how memory is the topology of the organism's routing of inputs (sensations) to outputs (behaviors).[50] Like a sieve or a filter, the organism's retention is inseparable from its system structure, its topology: it is a mapping of possible sensory inputs (stimuli) to possible reactions and reflexes. Just as in the three levels of retention described in Stiegler's extension to the bilevel Husserlian account, the protentions of one level operate a selection on the retentions of the level below it;[51] the overarching self-organizing system selects, through its own statistical sieve, those lower-system propositions that are better at identifying and retaining environmental features that afford the continued self-organization of the system as a whole. Thus, the whole constrains the potential expressions of the parts, which is the hallmark of emergent phenomena. A higher-order selection principle selects those individuals (among the parts) that selectively retain in such a way as to maintain holistic homeostasis, and thus sustain the metastability required of individuation. Memory is isomorphic to the system states "recalled" by such systems in their process of maintaining a dynamic equilibrium.[52] It is as though the system is constantly asking itself whether it has fully relaxed onto the attractor; and as long as it finds that it has not, it adjusts itself accordingly. Inference and conditionalization (reason) are thus implied by the thermodynamics of the extended critical system. It continually minimizes the discrepancy between the prediction and the event as a factor of the dissipation of excess stresses on its structures.

It is equally tempting to view each successive emergent level of organization, each new mnemonic exteriorization of the prior substrate, as resulting not from the pursuit of homeostatic stability through knowledge-informed inferences, but rather from an equally insatiable quest to experience what lies beyond the blind spots of perception: an explorative urge, unmotivated by self-preservation, to resolve what hides behind the horizons of thought. There is, as we have seen, a blind spot that necessarily appears in each retentional context's polarized sensory scape: though each critical system is defined *by* its polarized retentional constitution, it cannot itself directly experience that polarization, for it has nothing adequate to compare it to; there is no one-to-one correspondence or mapping between the operationally closed levels. Homeostasis depends on the adjustment of inferences to knowledge that is gathered about the world. But the evolutionary or technological exteriorization of the mnemonic organization, and its correlative redistribution across different retentional contexts (such as a mutation that changes a bodily structure or the spontaneous idea leading to the technological innovation), seems to depend less on selection and more on free variation and random exploration of the state-space liberated from the concern of utility. This happens, as we have seen, in all regions of underdetermination, wherever determinant factors neutralize each other, preparing the ground for spontaneous symmetry breakings.

This, again, reflects our discussion at the end of chapter 1: indistinct cognitions are correlated to series of operations that do not terminate. In other words, the experience of indistinction hinges on operations that do not suggest a means to recalibrate the boundary and minimize the discrepancy between the prediction and the event. The experience of the indistinct, in some sense, *demands* the exteriorization of the system. Gödel showed that the meaningful (true) expressions of a mathematical system exceeded what could be proven from within the system. But this also meant that proving those true expressions required recourse to a subsequent system, exterior to the first. Of course, this ultimately merely pushes the problem one step back, for the incompletion now reappears on the level of the new system, which now also allows true but unprovable statements. It is tempting to speculate, therefore, that the experience of indistinction is the *engine* behind the emergence of the successive levels of organism. When the self-organizing process discovers that some of its constructions do not terminate, the constraints of evolutionary selection, which require the system to *act before it is too late,* before it is deleted by evolution, motivate the production of a new subsequent system that encapsulates the first, affording those constructions a provisional terminus that successfully responds to environmental pressures. The exteriorization may result from the encounter between the organism's incompletion and the environment's selective pressures. This potentially allows us to situate indistinct perceptions, and indeed the aesthetic and artistic practices that cultivate them, as a functional part of evolutionary innovation. The experience of indistinction may be what motivates the expansion of the organism's phase space through new exteriorizations in order to provide new constraints that enable the completion of the constructions on the level below, while simultaneously revealing new incompletions and paradoxes on the level above.

In the thermodynamic interpretation of *fitness,* an organism is fit if its structure allows it to successfully dissipate the excess stress induced upon it by a changing environment. Evolving a flagellum will allow a hypothetical water-bound organism to swim away from its predator or find food. Evolving an eye will allow it to swim more efficiently (allowing it to descend to the depths of the sea to escape the ultraviolet light at the surface). It is this *plasticity,* this capacity to redefine its relation to its environment, that distinguishes the living system from other self-organizing structures, because, in effect, its capacity to dissipate stress affords the extension of its milieu or phase space. But it is also possible for an organism to become *too fit,* to become so narrowly adapted to its niche that it becomes vulnerable to changes in its greater environment. Hence, the system is better off, and more robust, if it injects some randomness into its process some percentage of the time and deviates from the course prescribed by the maximization of its expected model. This is precisely what provides biological evolution its remarkable robustness: through mutations and recombinations, it continually introduces variation into an otherwise closed system of prediction-error reduction. If it did not, life on earth would have ceased long ago, whenever the narrow circumstances affording its regenerative process would have changed. The discovery of incompletion (the experience of indistinction) and the subsequent supersession of local paradoxes through exteriorization is what maintains evolution's robustness in the face of change. Aesthesis and prosthesis are conceptually and functionally entangled.

These proto-technical exteriorizations, motivated by the indistinction and incompletion of the self-organizing system's limited experience of the world, are precursors to what we think of as essential traits of technical exteriorization. Recall that the rock does not care whether it melts, whereas the organism is defined by its striving to remain within a certain thermal phase. Along the same lines, the neuroscientist Karl Friston attempts to describe the brain and its mechanisms for learning and making inferences in terms of the thermodynamics of Bayesian evolutionary selection. Since organisms are embedded within the greater thermodynamic landscape, being (after all) complex dissipative structures, it follows that there should be some energetic principle that accounts for the emergence of these structures. We perhaps too hastily say that organisms are *negentropic*; in fact, they are not resisting time's arrow toward increased entropy so much as *delaying* this trend by, in effect, turning the arrow of time against itself to build structures that nevertheless still locally dissipate excess energetic stresses. If biological systems are structurally defined by how they selectively react to environmental cues in pursuit of this dissipation, there must be a way in which the thermodynamics of these systems gives rise to successful expectations, anticipations, inferences, or protentions. In support of this, Friston provides the following heuristic: though the crystal is a structure emerging as an efficient compromise between material and thermal constraints, it nevertheless can exist only within specific bounds and cannot self-adjust its relation to the environment in order to avoid the melting point. So, typically, snow eventually melts as it descends to lower, warmer altitudes. With this in mind, Friston asks us to imagine a snowflake that has wings and can self-regulate its altitude in order to perpetually remain cool, thus maintaining its characteristic shape.

> Here, we have taken a paradigm example of a non-biological self-organising system, namely a snowflake, and endowed it with wings that enable it to act on the environment. A normal snowflake will fall and encounter a phase-boundary, at which the environment's temperature will cause it to melt. Conversely, snowflakes that can maintain their altitude, and regulate their temperature, survive indefinitely with a qualitatively recognisable form. The key difference between the normal and adaptive snowflake is the ability to change their relationship with the environment and maintain thermodynamic homeostasis. [53]

Though this heuristic's imaginative use of a prototechnical exteriorization (the wings) as a means of avoiding the thermal threshold may convey the point Friston is making, it nevertheless fails to capture the *prosthetic, supplemental,* or *pharmacological* character of prototechnics. Following the logic of technics as prosthetics to the human body and the general principle of technical supplementation, on the level of the specific individual, something is *lost* in each successive level of innovation. A familiar example is that of our prehuman ancestors, who, as they evolved a large, thermally expensive brain, also needed to evolve a cooling system, such as bare skin and sweat glands, to offset the brain's heat production, and so they also had to lose the body hair. As in the Socratic doctrine of reminiscence, memory is *forgotten* with each reincarnation of the soul, as it is washed in the river *Lethe.* Similarly, each level of emergence can be thought of as a prosthetic supplement for the previous-level interactions that afford it; as Derrida stressed: "The supplement supplements. It adds only to replace."[54] In the case of our snowflake, this means that it could not grow wings while simultaneously *maintaining* its initial snow-crystal structure. That structure would have to be abandoned, replaced with a new structure allowing it to store information of a new kind. In other words, the wings could not be obtained for free, but would demand a *compensation* of the snowflake's structure. It would now need some way of sensing when the temperature is approaching a phase boundary, a minimal routing of an input stimulus to an output actuation. *Preferring* cold to hot implies some pragmatic knowledge of what "hot" *means,* some *memory* against which to compare incoming thermal stimuli. In this case, "too hot" means "flap wings." But this routing of a sensation to an actuation must be materially instantiated as knowledge or *memory.* It must be encoded somewhere, materially registered in the snowflake's variational *degrees of freedom.* And so, in order to have wings, the structure of the snowflake would have to be dramatically different, supplemented, reconfigured, and wholly organized around storing and recalling thermal information, such that it reacts appropriately to the looming phase transition.

The pharmacological and supplemental status attributable to technology, then, is not exclusive to tools, technical practices, and media. It is always already bound up in the very processes of cognition and life's extended criticality. Material habit formation in living systems, and perhaps some nonliving systems, is already *protomnemotechnical.* Habituation itself constitutes our experience of time as an accumulation and exteriorization of knowledge; the implicitation characteristic of habituation is analogous to the exteriorization characteristic of technology. This is what is meant by Stiegler's claim that "knowledge *is* technics,"[55] though perhaps what we are here touching on would be better rendered inversely: technics are extensions of the protomnemotechnical constitution of the organism, and epistemaesthetics thus inherently intersects the prosthetic logic of technics and prototechnics.

These observations suggest that humans are not so exceptional within nature as mnemonically exteriorized and retroactively interiorized aesthetic/prosthetic processes. For, certainly, each difference between one organism and another has everything to do with how they individually distribute themselves across retentional loci and their correlative economy of mnesis, amnesis, and anamnesis. What defines an aesthesic system is precisely its way of contracting events. So, whether it is completely isolated, contracting on one stratum only, or complex and distributed across many, this topology of retentional and protentional distributions is precisely what characterizes its own implementation and economy of aesthesis/prosthesis.

This again echoes our discussion concerning the entanglement between the supposed "two levels" typical of human cognition. There are, of course, many more than two strata of retention coalescing in the aesthesic integration of human experience, and yet so much philosophy has limited itself to describing but two relative levels: affective and discursive, implicit and explicit, or indistinct and distinct. For example, though there is a difference in *human* experience between what Whitehead called the *actual occasion* and the *event* (the molecule, his preferred example, is an objectification of multiple occasions), there remains the question of whether such distinctions should be ontologically reified or should be described as effects of perception and cognition. Whitehead's event seems amenable (though this is clearly not his intention) to the object of conscious *understanding*, a product of explicit reflective mentation (the *inferential claim*, the intentional act implied of our appreciation or objectivation of the molecule, to use his example), while the actual occasions seem to correspond to sensory input or the affective, sensuous level.[56] The same goes for the distinction between the different syntheses of time (Deleuze) or the distinction between retention and reproduction (Husserl). For, it should be stressed that, in complex organisms, each individual systemic level (each organic *tissue*, for example) is itself the location of an *evental adsorption*; insofar as it orients itself within time, it is temporally biased.

Thus, to the question of how far down we may say aesthesis goes, we might tentatively respond that it goes at least thus far: the pile of sand is already contracting events, and each successive, nested level of organization corresponds to an additional exteriorization of this simple principle. We humans are distributed mnemonically across diverse experiencing levels such as genes, cell membranes, nervous systems, cortical strata, social networks, and technological hypomnemata. But, perhaps because the cerebrocentricity of our human constitution "dominates" and "looks down" upon our other retentional/protentional loci, and furthermore centralizes our interactions with the transindividual realm of language and technology, we have tended to reify only "two levels": on the one hand, the sensuous, corresponding to a general indistinct mixing of all the retentions below, appearing as intensities and flux, and on the other, the distinct, explicit level corresponding to decisional operations on that indistinction.

The same can be said of the difference between sentience and sapience. According to some contemporary philosophers, though animals are sentient, they lack the sapience that characterizes humans. Sapience is sociolinguistic, they will argue: it is taken to be conditional on a network of reciprocally entangled commitments by various inferential claimants.[57] But what these observations suggest is that nonhuman sentience also requires such an entangled web of reticulations among inferential centers. Beyond the limit case of the "non-extended" self-organized critical system (such as the pile of sand), an event of cognition or recollection is always an instant of reciprocal operational satisficing among different retentional/protentional structures. This generalization is meant to naturalize my previous arguments against exceptionalist anthropocentric views on cognition and reason: languages reflect the same production of reciprocity on the level of human psychosocial transindividuation that is already driving the integration of the organism's sentience.

We see sapience as categorically different from sentience and knowledge as distinct from sensation only because of our biased normative human predicament. But the difference need not be categorical; rather, it may be said to be relative to the given organism's specific mnemonic distribution across retentional strata. Moreover, even among humans themselves, there is a plethora of different ways of becoming distributed across these retentional milieus (autism, schizophrenia, and other eccentric cognitive phenomena offer many examples, and different geohistorical technical environments and their variety of mnemonic affordances no doubt demonstrate additional instances of variance). As science-fiction narratives have oft explored, our potential transhuman and posthuman destinies also imply new ways of being distributed across these contexts, new hybridized forms of subjectivity.

This is precisely what is implied by David Roden's *disconnection thesis*, which I would argue applies equally well to any such incomplete mapping between cognitive systems with different retentional/protentional distributions, not just the impending disconnection between humans and posthumans.[58] One compelling example from the animal world is that there are different ways for an organism to be visually sensitive to frequencies of light: organisms may be bichromats, trichromats, tetrachromats, and pentachromats. There is no one-to-one correspondence between the respective color-spaces of organisms that see in three colors (as we do) and organisms that see in five (as with some birds): the topologies describing the links between retinal sensors in the eyes and the polychromatic cognitive "event" of vision cannot be mapped smoothly from one species to another, implying a "disconnection."[59] In precisely the same sense, due to the different possible ways a mnemonic critical being might metastabilize across a spectrum of retentional/protentional strata, the ways in which aesthesia takes shape in different kinds of organisms will most often be irreducible to each other.

A further point is that, insofar as they are essentially *critical*, temporally asymmetrical, we should regard each retentional substrate as already intrinsically aesthesic, in the sense that each concerns at least a minimal synthesis of events, or *integration of information.* The difference between levels of synthesis does not, therefore, necessarily imply an ontological distinction, but may more compellingly be referred to as a specific observation-bias-induced phenomenological distinction. Following the Copernican doctrine and the renewed calls to challenge anthropocentric thought in the Anthropocene period, this means we should refrain from ontologically reifying strict functional distinctions between actual occasions and events, or sentience and sapience, because those that immediately express themselves as the indubitable a priori foundations of thought may reflect only our own specific human metastabilization across retentional substrates. The epistemaesthetic domain necessarily extends beyond the mnemonic distribution that we psycho-socio-technical primates reify. Correspondingly, epistemaesthetics as a discipline must avoid falling into the trap of hypostatizing the human "faculties" while trying to understand their limited and provisional functional character within the larger landscape of non-human aesthesis.

4 SUPERPOSITION AND TIME

Hyperchaos and Superposition

There has been a revival of realist and ontological thought in the philosophy of the last decade. Perhaps as a backlash against the perceived stagnation of phenomenological, postmodern, and poststructuralist modes of philosophizing, which many theorists of the new generation have taken to be overly relativizing, we have seen the return of a demand for thought that does not shy away from addressing the question of the real, of the "great outdoors" beyond the limits of the phenomenal. The new debates between realist and relativist positions stirred by this resurgence necessarily intersect our mission to understand aesthesis and perceptronium. Indeed, the question of the possibility or impossibility of knowing the great outdoors is inseparable from the question of how objectivity and subjectivity fit into the greater *chaosmos*. So, now, after having observed that aesthesis is bound up with the mnesic synthesis of retentions and the organism's production of protentional horizons, let us examine how, as a process of transition between the great outdoors of the real and the correlation of objectivity and subjectivity, aesthesis is implicated in some of the central debates in cosmological and ontological theory of late.

Prominent among the various contemporary attacks on anthropocentric navel-gazing has been Quentin Meillassoux's very influential critique of "correlationism." In a bit over a decade since the publication of *Après la finitude* (2006), Meillassoux has spearheaded a wave of statements of decided positions concerning the status of realism, as versus epistemic relativism, in contemporary theory and was more than instrumental in bringing about the contemporary realist turn. Quite famously, Meillassoux accused the philosophical tradition since Kant of being antirealist, even solipsistic, in its reduction of reality to the *correlation* of thinking to that which is thought. Meillassoux remarks that, ever since Kant resituated the task of philosophy as that of revealing the conditions of cognition, and thus legislating between the knowable and the unknowable, things-in-themselves have been held to be inherently inaccessible, forever veiled behind appearances. He disparages how philosophy has strayed from its commitment to a reality existing independently of the observer. He argues that, through successive instantiations, from idealism to phenomenology, from pragmatism and constructivism to postmodern relativism, all philosophy has—implicitly or explicitly—become *solipsistic* since Kant. "Although these currents are all extraordinarily varied in themselves," he writes, "they all share . . . a more or less explicit decision: that there are no objects, no events, no laws, no beings which are not always already correlated with a point of view, with a subjective access."[1] Though individual forms of correlationism are complex and tedious to unpack, he maintains that the basic structure of correlationism is always the "minimal decision" behind anti-realisms. Against this obstinate trend, Meillassoux set out to say something about the real as it stands in the absolute, beyond the correlation.

He knew from the outset that his arguments against correlationism, no matter how meticulously performed, would be met with accusations of *performative contradiction.* He knew someone would always find the means to say that whatever is "out there" independently of us cannot be known *by definition,* for if anything could be said of things in themselves, we would immediately no longer be talking of things in themselves, but of things in their knowability, in their report to perception or experimentation. But Meillassoux's methodical argument turns that question around: is it not also a performative contradiction to believe in the scientific worldview, which implies an *ancestral time,* before the correlation? What he calls the *arche-fossil* is "a material indicating traces of 'ancestral' phenomena anterior even to the emergence of life."[2] He argues that correlationist ontologies are incapable of giving an account of the conditions of possibility of the arche-fossil, of things that existed before humans, before life, or before subjectivity of any kind. In other words, the arche-fossil concerns an appeal to matter as "a primordial ontological order: it is the fact that there must be something and not nothing,"[3] and the idea that mind or thought must have emerged from something that is not itself thought, that mind emerges from nonmind. Indeed, if everything known were a correlate of the act of knowing, then how could we say with any consistency that *we know* that the universe is approximately fourteen billion years old? This would indeed appear to be the correlationist's own performative contradiction. Meillassoux's quest against antirealism is therefore a defense of the scientific worldview, where the history of the universe, from the big bang to the present, and matter as the principle of the nonmental and presubjective are taken seriously and given precedence.

Meillassoux notes that, once the correlation had been realized by Kant, there was a "first wave" counteroffensive against correlationism that attempted to reinvent the notion of the real, not as that which was outside of thought, but as something absolutely congruent with thought itself: this he calls "subjectivism." The subjectivist will claim that the correlation *is* the "in-itself," that the circle *is* the absolute. Kant's first edition of the *Critique of Pure Reason* was accused of such a subjectivist idealism, from which he made a point of distancing himself in the second edition. But of course, this subjectivism became the founding motif of German idealism from Fichte to Hegel, where all that exists *is spirit,* where the real and the cosmos are simply the realizations of thought itself. But, if this idea eventually fell into disrepute, it is partly because the idea of the correlation *as absolute* was challenged by the scientific worldview, which revealed the *facticity of the correlation,* its merely contingent character. To show this, Meillassoux draws an analogy with death: "To think of myself as a mortal," he reminds us, "I must admit that death doesn't depend on my own thinking about death."[4] Here he is echoing the famous Epicurean argument, which many adolescents eventually discover for themselves: "Death does not concern us, because as long as we exist, death is not here. And when it does come, we no longer exist." But if we believe in the scientific worldview, and by consequence in a time that will come after our death, then, to be consistent, we must also assume that, when death comes, the correlation dissolves. Thus the correlation cannot be absolute; it is just a contingent fact; it *could* be otherwise.

Serving as an Archimedean point, it is this observation that allows Meillassoux to claim to find his way out of the correlation. It is "the very source which lends its power to the strategy of de-absolutization" of the correlation that, ironically, "*also furnishes the means of access to an absolute being*" outside the correlation.[5] Because "the non-being of the correlation" is conceivable, this very conceivability opens a window onto the facticity that subtends the correlation. Meillassoux arrives at the following assertion. We *can* say something about the great outdoors, about what there is beyond the correlation, and we can know something that remains within our knowledge of it just as it is beyond our knowledge of it, beyond thought's conditioning of its independent existence: the absolute. And it is the very idea that the correlation is itself merely a contingent fact, and therefore could be otherwise, or could *not have been the case,* that allows Meillassoux to say that there is indeed a knowable thing outside the correlation. This radical contingency behind the correlation is itself not part of the correlation; it is outside, and yet knowable as such.

Meillassoux then makes a subsequent move: in his consideration of this facticity behind the correlation as an absolute, he realizes that it constitutes a form of contingency more radical than what we usually think of as contingency. Indeed, contingency usually has two aspects: it means both *not necessary* and *depending on other factors.* But Meillassoux's radical notion of contingency drops the second sense of the word. The kind of contingency behind the correlation is not dependent on anything else. He calls this *hyperchaos,* "the absolute absence of reason for any reality, . . . the effective ability for every determined entity, whether it is an event, a thing, or a law, to appear and disappear with no reason for its being or non-being." Hyperchaos is a kind of *time without becoming;* it can change the state of the world without warning from "outside" of time. He thus comes to advocate a position that, at any given moment, anything and everything can be radically modified without reason, without cause or dependence on anything else, and thus without being deterministically bound to any state of affairs. The laws of physics can simply switch to other orders, the symmetries and invariances we observe in the universe hold to absolutely nothing, and impossible objects can pop out of the void. Even God, he claims, can one day rise from inexistence. Meillassoux therefore has to distinguish this hyperchaos from what is known as deterministic chaos: where the chaos we normally refer to happens in time, this happens outside of time, and thus it is, as he says, "time without becoming." Here (and this will become important for us later on) he is remarkably close to Gilles Deleuze's arguments for the "static genesis" that conditions causality and deterministic time. Indeed, for Meillassoux, the notion of becoming is sterile: becoming merely follows the strictures of the causal relation, whereas hyperchaos is the truly creative principle. Correspondingly, he argues that Heraclitus is a "terrible fixist": for if everything is constantly changing, then, in a sense, nothing is changing, and the world just flows along like a river, always the *same* river. (Consider again the case of Clive Wearing, who, having lost the capacity to create new episodic and semantic memories, is trapped in this Heraclitean flux, where each moment is always and forever new and original, but is always in some sense the same moment, devoid of individual character, for it cannot be compared to any other moment different from itself.) Hyperchaos, by contrast with Heraclitean flux, means that the canonical river can at any point spontaneously be replaced, without warning and without reason, with a mountain or a hypercube.

This argument has certainly caused a stir. The movement of loosely related speculative philosophies associated with the term "speculative realism," despite their divergent ontological commitments, are said to have agreed on one common motive: they were united in their cause against correlationism. The Meillassoux effect was perhaps a black swan, emerging spontaneously from nowhere but creating waves through the philosophical blogosphere, where an abundance of young speculators eagerly sought strategies to *disrupt* the academically sanctioned strictures of philosophy. A generation of philosophers demanding more elbow room in a sedimented and constrained world of knowledge could now, thanks to Meillassoux, unanimously accuse philosophy of a great performative contradiction: how can you believe in science and yet be a correlationist?

Michel Bitbol has answers to this question. He could well be perceived as one of these settled academics that the anticorrelationist disrupter aspires to displace. He is precisely the target of such accusations: he defends science at its most fundamental level, and yet is a card-carrying correlationist. Indeed it would be difficult to find a philosopher more at odds with Meillassoux's position: Bitbol, an erudite and established French philosopher of science, has staunchly defended a radical correlationism since long before the term was being used. Being a Kantian, he understands the task of philosophy as that of revealing the conditions of possibility of knowledge. His philosophy therefore advocates a radical transcendental relativism that allows no form of absolute, no certainty about anything outside the correlation. And yet, Bitbol insists on the importance of *metaphysics*: it is the science of *autoreflection,* of situating the act of knowing within the process of cognition itself to discover the boundary conditions of science, of knowledge, and even of existential questions. His stance is persistently autoreflective: to be philosophically critical is not only to know, but to know one's knowing, to observe the act of observation in the making.

At a meeting in Paris on the new realisms where Meillassoux and Bitbol were presenting their divergent views on correlationism, I had the chance to ask Bitbol the following question: what do you make of what Meillassoux calls the "ancestral"? What, in your view, came *before* the correlation, before the first living cell or the first autopoietic unit, whatever it may have been? Bitbol's answer resonated with me for months: before the correlation, he responded, "there could only have been a quantum superposition of states, a pure indeterminacy."[6]

His remarkable response seemed to suggest that, in his view, the past may have been "printed out" retroactively from the original autopoietic or correlational entity, as though the ancestral realm and the archefossil were secreted progressively in the correlation's successive cycles. This response was perfectly in line with Bitbol's larger research agenda, which is deeply influenced by Kant, Husserl, and the revelations of quantum physics. He is known in philosophy of science for the rigorous transcendental interpretations he has advanced describing various aspects of quantum physics. According to Schrödinger's equation, the cat is both dead *and* alive before one takes a peek inside the box; that is, the wave function or the state vector defines a system *in superposition*. In the traditional view, the wave function "collapses" onto one state or another, choosing, for no observable reason, one or the other of the states previously in superposition. In his view, quantum phenomena do not describe the real, but rather reveal the *boundaries of experimental knowledge*. In this assertion, he is following Kant's insights quite closely, save for one very important modification inherited from Hans Reichenbach and Ernst Cassirer that we have encountered previously: that the Kantian a priori must be *relativized*. In other words, one cannot stipulate universal a priori conditions for all acts of knowing, because they differ in relation to the object of knowledge being apprehended. We touched on this in chapter 1 when we looked at how the discovery of non-Euclidean geometry and Einstein's related theories of relativity seemed to put into question the Kantian project. The neo-Kantians responded by relativizing the a priori to account for the fact that different a prioris will participate in the constitution of different objects of knowledge, but stressing that, in each case, the a prioris remain necessary correlates, implicit conditions of the act of knowing. Bitbol sees this reflected in quantum physics, where the experimental context always brings with it certain a prioris, implicit assumptions that structure the problem and are, thus, coconstitutive of the object observed. The wave function and the state vector, as ways of describing the phenomenon, are also constitutive in this way of the object being observed, which is why the states of the observer and of the measurement apparatus are inseparable from the quantum-mechanical system being observed.

For this reason, Bitbol is quick to insist that the ancestral superposition has "no preferred basis," that it cannot have been the superposition specified by any of the particular modeling contexts we use for describing the quantum reality of the universe, such as the Hilbert Space. Likewise, Bitbol maintains that we should not be "realists" about the wave function or the state vector, because, "in practice, a state vector is little more than a mathematical tool for computing probabilities of experimental events,"[7] and that the superposed states described by such models are therefore conditional on the specific experimental setup used for measurement. "For the state vectors to manifest a specific [*définie*] propensional structure, that is to say, for them to indicate the tendencies of a system to manifest itself in phenomena of a given range [*gamme déterminé*], they must in effect be decomposed according to the base of a certain observable."[8] He thus argues against the Everettian (or many-worlds) interpretation of quantum mechanics and the decoherence theories that derive from it, which understand the *real* thought-independent and *absolute* nature of the universe as being a monstrous superposition of such states. In true neo-Kantian spirit, he contends that the experimental apparatus and the model used for describing the nature of the system are *constitutive* of the possible states of the system in question. The superposed states are therefore *correlates of the experiment,* or of the a priori conditions of the act of measurement. In other words, the superposition does not correspond to the *real* so much as it reflects the boundaries of the experimental context. No amount of combining or multiplying of these possible superpositions, he argues, will result in any certainty of congruence with the noumenal real, for they will always be extrapolations of our experimental orientation within the world.

Correspondingly, Bitbol stresses that even his own speculation as to the nature of the great outdoors still does not escape the trap of correlation: he has merely abstracted the correlate to the second degree, twice subtracting the relation between observer and world. In other words, the stipulated great outdoors, even with the extra clause of being "without preferred basis," remains relative to the act of observation, and thus, he argues, no matter what we come up with as a characterization of the world beyond correlation, we are merely attempting to subtract the relation from the act of observation; we are not really touching or knowing the observer-independent real.

Meillassoux's and Bitbol's respective positions are therefore diametrically opposed. Meillassoux claims to break out of the correlational circle and access the great outdoors, while Bitbol argues that there is logically no way to do this, *by definition*, and that the point of philosophy is not to understand what there is outside the correlation, but rather what the conditions of this correlation are and how they evolve.

And yet there is nevertheless a kind of symmetry between the two positions. For, as we have seen, what Meillassoux achieves with his meticulous argument is, in the end, very similar to what Bitbol (hesitatingly) speculates must logically precede the epistemic correlation. For Meillassoux, the observer-independent real is a hyperchaos, which he also characterizes as *time without becoming*. For Bitbol, the real outside is a pure superposition without basis, *before* time. Notice that, in this characterization of the "absolute" outside, Bitbol's quantum picture resembles the hyperchaos Meillassoux describes. So, quite ironically, the correlationist and the anticorrelatonist defend nearly assimilable views. The primary difference here is that the correlationist claims that the outside is intrinsically *unknowable* because it is, by definition, the indiscriminate superpositon of that which *could* be known, while the anticorrelationist claims that the real *can* be known, for insofar as we can deduce that it is some kind of indeterminacy (chaos or superposition), we are indeed in possession of some knowledge of the outside. Thus, it would seem that the difference between the perspectives is a mere matter of epistemological emphasis.

But one more thing can be gleaned from the comparison. It is furthermore ironic to note that, in a sense, in his cautious response to my question, the correlationist Bitbol seems to obtain something even more removed from the correlation than does the anticorrelationist Meillassoux. For, a hyperchaos is not exactly an atemporal superposition without preferred basis: a hyperchaos, expressed as a *time without becoming*, is nevertheless a kind of time, since it *changes* from one state to another, despite following no deterministic laws. In other words, rather than consisting of an absolute symmetric superposition of possible states, its *temporal* expression implies that it is already an *asymmetrical sequencing* of those states, in the sense that the states are serially iterated, even if purely randomly. Each iterated state is exactly congruent with the *privileging* or *selecting* of one correlational point of view (one act of measurement) or another, which comes with its own implicit assumptions or a priori conditions. Thus, in a sense, it is more subjective, more correlational, to posit the great outdoors as a hyperchaos than it is to posit it as a great atemporal superposition of superpositions.

Regardless of whether we hold that our speculation about the observer-independent reality is knowledge or not, what is clearly exposed by this approximate symmetry between the two positions is that whatever is posited to be outside the correlation will tend to be symmetrical and indiscriminate with regard to the discriminated asymmetries of the observed world. To say that what exists outside the correlation is hyperchaos or hyperposition testifies to the issue that, when looking for ultimate causes or explanations for *why* causality exists in the first place, or again for why *there is something rather than nothing*, the explanation can have but one possible character: some sort of spontaneous break in symmetry, some uncaused bifurcation, a swerve from eternal silence.

Following Solomon Maimon, who attacked Kant's project for not being able to give an account of the *genesis* of the conditions of knowledge, of the formation of a subjective experience from a nonexperiential background, we can similarly draw attention to the fact that these speculations on what is outside the correlation do little to describe how this relation came to be. In this regard, Meillassoux's *sequentialization* of the static hyperposition of possible states as time without becoming, as a random sequence running through the possibilities in time, perhaps gets closer to responding to the question of the process that leads from the precorrelation to the correlation. And yet, as we will see, the process itself remains unsatisfactorily flattened onto the subject–object relation. Bitbol, for his part, seems unwilling to venture into this speculative territory: he wavers between a perspective according to which the correlation must be stipulated from the outset and a perspective in which the correlation emerges spontaneously from the indiscriminate undetermined background. But the question of aesthesis is precisely the question of how these conditions arise, which is why we must take a closer look at the problem of the *conditions of conditions,* the *causes of causality,* as well as at the abstract process that leads from symmetry to asymmetry, or from probability to improbability.

Causality and *Quasi-Causality*

The discrepancy and simultaneous quasi symmetry between the positions of Meillassoux and Bitbol can be regarded as a contemporary echo of a very ancient dilemma. Aristotle realized that a problem of infinite regress challenged the consistency of all knowledge:

> [If] there is no way of knowing other than by demonstration, . . . an infinite regress is involved, on the ground that if behind the prior stands no primary, we could not know the posterior through the prior (. . . for one cannot traverse an infinite series): if on the other hand . . . the series terminates and there are primary premises, . . . these are unknowable because incapable of demonstration.[9]

Aristotle rejected both alternatives and proposed that there was a kind of knowledge that was valid a priori, self-demonstrably, and that all subsequent "demonstrative knowledge must rest on [such] necessary basic truths."[10] The distinction between those kinds of knowledge that are self-explanatory and those that rely on references to other knowledge would become the distinction between necessary and contingent truths, which would again be echoed in the discrepancy between analytic and synthetic: necessary truths are self-substantiating, a priori, whereas contingent ones depend on and refer to others. In his cosmological argument for a universal *sufficient reason,* Leibniz made a similar observation: "The sufficient or ultimate reason must needs be outside of the sequence or series of these details of contingencies, however infinite they may be."[11] On Leibniz's account, there must be a cause that is *outside* of time, outside of the contingencies of efficient causality. The reason for contingency is not itself contingent, but necessary: *a truth that is its own cause*; a causal loop outside of time. For Leibniz, explaining this involved two variants of his principle of inclusion: in the case of contingent truths, the subject includes the predicate unilaterally, whereas in necessary truths, the subject and predicate reciprocally include each other.

This ultimate self-consistent truth, in addition to obviously intersecting the problem of foundationalism, foreshadowing arguments like Wilfrid Sellars's "myth of the given" and still looming as the only potential way out of the perceived arbitrariness and untenable relativity of coherentist theories of truth, also corresponds to the cosmologist's dream of discovering a unification of fundamental physics as a "theory of everything." The American physicist Steven Weinberg, a prominent defender of the "dream of a final theory," claimed that "it is very difficult to conceive of a regression of more and more fundamental theories becoming steadily simpler and more unified, without the arrows of explanation having to converge somewhere."[12] Against the position held by those—for instance,

Karl Popper—who deny that there could be such a thing as an "ultimate explanation," Weinberg argued that it was unlikely that the chains of explanation could go on indefinitely with no end.

Interestingly, at the opposite end of the spectrum of thinking on this issue was Weinberg's mentor and friend, John Archibald Wheeler, who arrived at a somewhat different conclusion. He recalled the mythical story of the world-bearing turtle, according to which the earth stood on the backs of elephants, which in turn stood on the shell of a giant turtle. But of course the myth seems to beg the question: what was the turtle standing on? Is it turtles all the way down? Rejecting such an infinite regress, Wheeler argued as many before him that one must posit a loop of some kind:

> To endlessness no alternative is evident but loop, such a loop as this: Physics gives rise to observer-participancy; observer-participancy gives rise to information; and information gives rise to physics.[13]

There is a helpful way to construe what Wheeler was putting a finger on: the *matter-information-observation loop.* A consequence of the quantum physical entanglement of observer and world is how *matter* reveals itself *as information.* Wheeler understood that quantum physics exposed the informational character of science's disclosure of nature. Everything we call "nature" reveals itself through specific perturbations and actuations by our organism and its technical prosthetics. Nature emerges forth only as equipment-evoked responses to yes-or-no questions: in each case, what is known about nature takes the form of a *yes* or a *no,* a this or a that, a here or a there, a now or a then, a 0 or a 1, and so on. "It from bit," as Wheeler put it. In each case, nature's response comes in the form of information.

But, according to Claude Shannon's equation, information is nothing but a measure of the observation's unexpectedness, or, in other words, a measure of the improbability of the event given the observer's prior conditionals or expectations. It derives from the question "what may we expect?" or "what can we predict?" For instance, in the canonical example of electrical signal communication, information can be encoded prior to transmission by modulating a carrier wave (recall the familiar terms *FM,* for "frequency modulation," and *AM,* for "amplitude modulation") and can then be retrieved again at the point of reception by *subtracting* the carrier from the communicated signal: information generally corresponds to the amount of meaningful *difference* between what is expected and what happens to be experienced. But in that case, if information derives from observation, what then is observation? What is the status of the subjective perspective in its relation to world?

Here things exhibit the typical quantum strangeness, for science is forced to say that observation too is "made of matter," and thus that it too is ultimately made of information. The observation itself is part of the universe, as is the observed. Thus, what is mind other than an exotic state of matter, a matter that has experiences and retentions, a matter that is critical and selective and conditionalizes its protentional outlook on prior experiences, thereby participating in the very constitution of nature as information, and a matter that is nevertheless ultimately composed of information itself? But how can information *depend* on observation while also being its *precondition*? For, on the one hand, observation derives from the contingent toils of matter, consciousness emerges from a nonsentient inorganic material, and life emerges from nonlife, but, on the other hand, at the same time, all matter is information, and all information *derives* from observation. In other words, matter is derived from information (symmetry breaking in the early universe), information is derived from observation (for, it is founded on a probabilistic or predictive scheme), and observation is derived from the contingent toils of matter (for, life emerges from nonlife), and the cycle begins anew.[14]

But again, though seemingly diametrically opposed, Weinberg's and Wheeler's positions merely emphasize different aspects of the same problem. Weinberg is saying that we will arrive at a theory for which, like Leibniz's necessary truths, there will be a reciprocal inclusion between the subject and the predicate, while Wheeler is saying that the chain of explanations will, in the end, loop back to the beginning. Both positions carve out their own special distribution of *the line and the circle*: the line of the explanation, the circle of the self-relation. Wheeler's wager against infinite regress leans toward Bitbol's Kantian interpretation of quantum physics: no matter how far back we peer into the cosmos, or how far we extrapolate the current asymmetrical situation, or how good our models of the universe become, there is an infinite regress toward a supposedly symmetrical origin. Thus, we are invited to situate the real source of knowledge in the relational loop itself, while also leaving a symmetrical negative space for positing an always already absent original symmetry, *before* the first break. Therefore, both Wheeler and Weinberg, in their respective ways, entertain the concerns of Leibniz and Aristotle: if we want to avoid the infinite regress, we have to posit a loop of some kind. Even Meillassoux's account, [15] which, on the face of it, appears to defeat the circle, seems to get us no further: his claim that the only necessity is contingency itself can be said to replace the logical circularity of the human–world relation (correlationism) with that of the hyperchaotic relation of the *absolute with itself.* In order to break the correlation, he is forced to induce an essential unrest in the depths of matter, a necessarily contingent origin that can be seen as an operation of self-positioning, a circular reference of the static superposition to itself, giving rise to its sequential yet random expression. But, though more subtle, this rendering still seems to account for the line of explanation, entailment, or causality only through a tautological loop.

This distribution of line and circle seems to intersect epistemaesthetics at all levels of engagement. Indeed, if aesthetics as a discipline puts so much emphasis on the notion of *judgment* (the judgment of taste), it is in part because of the subjective or biased character of the "universal assent," the "ought to," that Kant rightly recognized was the peculiar mode of being of the organism in its self-organization. Hume had observed that there was no necessity other than that which emerges from the subjective habit of linking the past to the future, the cause to the effect. And to judge, as Kant noticed, is always to decide in the place of others, and thus to pretend to the universal assent of a subjective impression via the curious operation of "subjective necessity." Kant already understood that there was a natural purposiveness founded in the organism's constitutive structure, corresponding to a local necessity: the organism's pursuit of continued existence. In other words, this necessity was the subjective correlate of the objective or natural purpose that characterized the living being, for which "the parts of the thing combine into the unity of the whole because they are reciprocally cause and effect of their form." [16] In his *Naturphilosophie,* Schelling went on to describe how this correlated autopoietic being, a *natural monad,* a kind of "whirlpool" of material flows, may have arisen from the nonautopoietic material background. At its most abstract, aesthesis can be nothing other than this very process: the production of a subjective closure from an asubjective and open material origin. It corresponds to a skewing and warping of matter that renders it critical, selective, and poised. It therefore also concerns the transition from contingency to necessity, for a relative *finality* or *telos* emerges only once this criticality has set in. Indeed, the emergence of the natural monad is simultaneously both the inception of time's arrow (this is why, as we saw above, Meillassoux's characterization of hyperchaos as a time without becoming remains explicitly subject-relative) and space as distinction between here and there, local and global.

We are faced with a comparable assessment if we consider things from the perspective of Bitbol, who radicalizes and relativizes Kant's position: necessity is *necessarily contingent,* for if one considers the universe from the point of view of quantum physics, where observer-independence implies an indiscriminate superposition of all possible worlds, all possible observations, then there is no sense in positing necessity outside of specific and contingent correlations between subjectivity and world. Before or beyond the correlation, it seems there can be only an absolute indiscriminate mixture of time and space; bodies, states, configurations, and such are smeared out across multiple dimensions, infinite arrays of potential attributes, effectively dissolving their cogency. A necessity automatically holds to a certain perspective on things. A necessity emerges once a system has evolved, purely contingently, to some critical state and once there emerges a first distinction between self (defined as that which is continually regenerating its constitution) and nonself. Kant can be said to have recognized this strange reversal of priority between necessity and contingency and begun to describe the deep contingent background that results in necessities. This is indeed how Deleuze reads the significance of the Third Critique. Rather than coming *third* to understanding and morality, the aesthetic is not itself a faculty, but "provides a basis" for or "makes possible" the faculties: it configures the ground on which they take hold, where their legislative authority obtain their license. Thus, aesthesis must be rethought, not as a lower level of cognition, but as the very formation of the principles that allow subjectivity and objectivity to take hold through an economy of circles and lines.

This is perhaps why science is forced to, in a sense, operate in the contrary direction. All knowledge can be said to climb progressively back up the chains of causality (or explanation), which never cease to move from order to randomness, from improbable to probable. The universe, we say, is *relaxing* onto its most probable state, and this relaxation corresponds to the second law of thermodynamics. As the universe decomposes into randomness, it scrambles its origins, and science is the business of deciphering just how the necessities have been progressively obscured by these interferences and mixings (entropy). And so, we are faced with a world that is contingent and mixed, and our knowledge seeks the essences, the necessities behind appearances. Science asks what the underlying principles that structure and constrain the events we observe are. What are the necessary truths behind contingent phenomena? The causal chains physics must unravel, therefore, inevitably flow from the improbable to the probable. Indeed, the probable is the great attractor of all causality. Under the causal constraint of free energy reduction, the system is dominated by the principle of least action, which is why Humpty Dumpty cannot be put together again after his great fall. Causality itself cannot be distinguished from the very process of descent onto the system's most probable configuration, the attractor state, which is why all scientific explanations are therefore appeals to the relatively improbable states that preceded some phenomenon in question, and for which the phenomenon was the probable resolution. The *absolutely* probable finality of all this material relaxation corresponds to a statistical symmetry, which is why randomness, as such, is *that which requires no explanation.* As in the Hermann von Helmholtz theory of *unconscious inference,* we perceive only configurations in our environment that are relatively unlikely or improbable. No one ever has the impulse to ask why such and such a thing is randomly distributed, why a given system is at equilibrium, or again, why two events appear to be unrelated? On the contrary, it is always contingent correlations and dependencies between occasions of experience that motivate an initial interest, involvement, or concern for the world: the signal in the noise, the curious pattern, the conspicuous discrepancy, and the covariant behavior. In other words, because they are less probable, only deviations from randomness will beg for scientific explanation.

Recall that, as we have seen, the organism cognizes changes in the environment through its passive contraction of events. It therefore registers that which is not at equilibrium, that which still maintains enough resilience and recalcitrance to not be muddled with everything else. This is why, in Shannon's theory of information, uncertainty reaches a maximum in the equiprobability of all possible events: the entropy is highest when the probability distribution of the given system is uniform. The lowest energy, and therefore the most probable state of the communication channel, is not the bare carrier wave, which is already quite improbable but always ultimately pure noise, an absolute absence of signal (or an absolute superposition of all signals, which amounts to the same thing), a uniformity of the distribution. In the causal regime, which is always already fully actualized and coherently implied by the facticity of the world at hand, the process evolves from the least probable to the most probable state. This implacable movement, of course, corresponds to the second law of thermodynamics: the arrow of time. Signals, patterns, and significant features will eventually fall back into the randomness of the noise floor as the system dissipates its energy and evolves toward the highest entropy state, for causality is dominated by the *principle of least action.* And the thermodynamic arrow of time furthermore echoes the *arrow of logical priority* inherited from Aristotle and Locke. Necessary properties are primary and logically prior to accidental properties. A square's four sides are *necessary,* essential properties of the square and "come before" any of its contingent, accidental, or existential properties—for example, its color. The *red square* could not exist as such if it did not first have the essential properties of squareness, without which it would not be a square of any color whatsoever. This is because the subjective purposiveness that conditions any demand for an explanation *always comes too late,* and is thus condemned to climb back up the chains of influence, whether it be causality's movement from improbable to probable or logic's priority of necessary over contingent.

This all seems plainly obvious. But our account of aesthesis has corresponded to a different movement. That is to say, to explain the subjective necessity that grounds the question itself, to explain the asymmetry of time itself, we are forced to appeal to a somewhat more elusive process in the reverse direction. There are, therefore, not one, but two distinct forms of dynamism to account for. There is the well-known process that leads from necessity to contingency, or from improbable to probable, and we call these "entailment" and "causality," respectively. But there is a second process to speak of, a "quasi-causal" process, to enlist Deleuze's concept, that operates in the reverse direction: it produces necessity from a contingent source. It is this that allows for Humpty Dumpty to have been sitting on the wall in the first place. Richard Feynman once put it thus:

> So far as we know, all the fundamental laws of physics, like Newton's equations, are reversible. Then where does irreversibility come from? It comes from order going to disorder, but we do not understand this until we know the origin of the order. Why is it that the situations we find ourselves in every day are always out of equilibrium?[17]

Indeed, why is the world not already at equilibrium? Humpty Dumpty *could* have started off on the ground in the first place. That would have been much more probable, and indeed, the only recourse for science is to extend the chains of causality even further back and posit an earlier, even more improbable and even more ordered state before Humpty Dumpty, before the big bang. Like a curious child, we can always repeatedly ask, *but why?* Like the tortoise in Lewis Carroll's "What the Tortoise Said to Achilles," we can always demand of Achilles an additional condition that would justify the proposition at hand.[18] Likewise, the past is in effect the perpetual extension of the chain of conditions explaining the present. So, as long as we are trying to explain a system's evolution in time, the past can always have been only in a prior, less probable state than the relatively more probable posterior state. The *debt* owed by the prior to the posterior therefore had to be repaid, compensated *in, through,* or *with time.* Time is compensation for the improbability of the origin.

Very importantly, however, in contrast to these characteristics of causal time, Deleuze conceives of *quasi-causality* as evolving from the probable to the improbable. It thus engages with this business of why there was order in the first place, why the universe, for the observer, is in a relatively low entropy state at the beginning. Going against the principle of least action (and hence challenging the purview of Occam's razor) *quasi-causality* is the spontaneous production of asymmetry, corresponding to what in cosmology is called "spontaneous symmetry breaking." For, in *quasi-causality,* necessity emerges extemporaneously from randomness; symmetry is spontaneously broken. Ultimately, this spontaneous break in symmetry, this contingent leap into necessity, can be construed as the *logical condition* of causality. As the temporal or processual being of that which results in causal regularity and determinism, it is congruent with Meillassoux's notion of time without becoming, as well as Deleuze's notion of *static genesis.*

Of course, from the point of view of causality, this process does not make sense. Why would and how could anything go against the principle of least action? Again, this reverse conception of causality seems to imply a circularity. It is as though the new asymmetry *retroactively* constitutes its origin, its *quasi-cause.* In other words, the emergence of subjective necessity seems to spontaneously, and acausally, induce its contingent origin in the *future perfect* tense. But, in order for Ouroboros to stop chasing its tail, in order to escape the banality of the assertion that the observer is always correlated with that which is observed, we must consider another aspect of the issue: how *quasi-causality* is bound up with the complexity and variety of the seemingly irreducible entities populating our world.

From the point of view of fundamental physics, there would seem to be only one reality. There are no stratifications between regimes: no distinction between biological and chemical, no leap from the subatomic to the atomic, no difference in kind between mind and matter. In fundamental physics, everything is information. From this perspective, the microphysical reality absorbs all other accounts, since it is unchanged throughout the toils of the macroscopic world. As we have seen, we can take the position that all of this information ultimately reduces to a self-consistent formula, a self-positioning of the absolute with itself, or we can take the position that all this information points back to the fact of observation. Between the circle of subjective necessity and the circle of objective necessity, there is an entire world that unwinds, carving out its own asymmetrical existence from the otherwise absolutely reducible cosmos. The main challenge to "dreams of a final theory" is that those doing the theorizing (the thinkers, knowers, or observers) do not live in the ideality of the world perceived by an omniscient demon. This is due to an important Leibnizian discovery. We do not live in pure possibility: a world is always composed not merely of that which is possible, but of that which is *com-possible*. A world is not the sum of possible things, for indeed, some possible things, though noncontradictory, would nevertheless be incompatible with the other things populating the world. And so, if science is an extension of our organismic compulsion to make better and better predictions, it needs to account *not only* for the worlds that *could* have been *but also* the world that factually happens to be the case, the world in which this history, this scientific culture, and these observations take place. As soon as this is admitted, it becomes clear that nature, insofar as it is observed, cognized, or experienced, can never completely unfold into the perfect distinction and clarity of a mathematical formula or final theory of everything. Nature is irreducibly composed of many stratified layers of intelligibility, which is why chemical phenomena cannot be neatly reduced to quantum phenomena, and why biology cannot be appropriately described solely on the basis of chemical reactions.

These stratifications and coarse-grainings of nature are inseparable from the path-dependent character of our organismic past and its process of cognition. They subsist as the materiality of the transcendental conditions of our forms of subjectivity. We may have a theory describing the rules of efficient causation between these regimes, and we may establish complex symmetries that describe *possible* events occurring as the result of observed phenomena, but we will never get an overall view of the *compossibility* of the events populating our world. In reality, when it is time to specify the individual cause of an individual effect, we find that the causal roots of all but the most trivial events are irrecoverably scrambled in the past, precluding our capacity to obtain an explanation or prediction, save for mere *possibilities* (the imprecision of which is correlative to the coarse-grainedness of our observations). And indeed, as Henri Bergson argued, possibilities are products of the actual: they therefore do not precede the scientific attitude's modeling of the problem, but in fact emerge from the model's structuring of intelligibility.

The last century's debates on the nature of emergence, irreducibility, downward causation, and nonlinearity all turned on this question: is irreducibility merely an effect of the observer's failure to discern the minutiae and its imperfect resolution of the details? Are chaotic phenomena byproducts of the observer's blind spots that invariably cause them to overlook the microstates of the system? Is time merely an illusion? This is no doubt the orthodox contemporary view: since Einstein, we have understood time as a dimension of space, a function of a body's acceleration in spacetime relative to other bodies. But of course, in general relativity, time is perfectly reversible. As Einstein famously quipped, "The only reason for time is that everything doesn't happen at once." And indeed, objectively speaking, from the point of view of a hypothetical omniscient observer, everything *would* seem to happen at once. Recall that, for Laplace's demon, who could perceive in perfect *distinction* all the causes of all the effects in the universe (that is, who could see how all the events populating time were deterministically connected with each other), "nothing would be uncertain and the future, as the past, would be present to its eyes."[19] But this is the limit at which the metaphor breaks down, for in another sense, for the omniscient observer, nothing would *happen* at all. A happening requires an extra element. The mere togetherness of the elements involved is not, it would seem, sufficient to account for the event's happening. Nor, indeed, is it sufficient to account for the observer, for what is the observer without the observation?

This consideration leads us to the question of the reality of determinism and monism. If time is simply a sequence of deterministically interconnected "events," we are left with a picture of time that is equivalent to timelessness: everything has always already happened; the future is already "written." In other words, regarding time as a purely deterministic irreversibility does not allow us to escape the prospect of cosmic timelessness. Thus, a true realism of time requires that the future be conceived as open. William James argued that confidence in monistic determinism required that we relinquish all claims to chance and free will. He quite rightly acknowledged that we could not objectively decide whether the world was at base a deterministic monism or an *indeterministic* (or partially deterministic) pluralism. However, using an argument reminiscent of Pascal's wager, he argued that the assumption of pluralism and partial indeterminism was a better wager, a more pragmatic stance to adopt, as it allowed the possibility of chance and volition.

More recently, Roberto Unger and Lee Smolin have been reviving such arguments against the scientific denial of time and history. Together they claim that science has been swept up by a misguided mathematization of the universe that has confused the model for the cosmos itself, and has therefore progressively also abandoned all logical grounds for political agency, accountability, subjective awareness, and perspective. They make a compelling case for radically rethinking cosmological physics: instead of positing a *timeless multiverse,* where every event and its opposite actually and "eternally" exist, they argue that we should commit science to the idea of a *singular universe* that is everywhere affected by *open-ended change* and to the humility of understanding that a model is always provisional and revisable; it is just a model, never the universe itself. All is affected by time, they argue, even the *ways* in which the universe changes. Instead of merely multiplying dimensions and branches of the multiverse, we regain some grounds for ethics and politics (and, I might add, aesthetics) by realigning cosmology with the idea of a one true reality "drenched" in time.

> If our universe were only one of many, inaccessible universes, mathematics could form part of the science of the timeless totality of these remote and hypothetical worlds. However, if all we have is this world in which we awake, and nothing remains outside of time, we can give ourselves—or mathematics—no such excuse. [20]

They argue that cosmology's tendency to pluralize the universe and trivialize time, to pulverize possible worlds and diminish the importance of historical analysis, actually goes against the central doctrine of scientific reasoning, meaning Leibniz's principle of sufficient reason. They remind us that, as Charles Sanders Peirce noted, nothing is in so need of explanation as a law.[21] Smolin in particular claims that the principle of sufficient reason should be regarded as insisting that there should be "no ideal elements or background structures in the formulation of a truly cosmological theory."[22] The principle of sufficient reason, in other words, rejects all mathematical objects or structures that are "specified for all time, have no dynamics, participate in no interactions, but are necessary to give meaning to the degrees of freedom that are dynamical."[23] In this regard, Smolin notes, there is no difference between the Hilbert space used in quantum mechanics and the "absolute space and time" stipulated as a presupposition in Newton's physics, as well as, I might add, in Kant's metaphysics (the transcendental aesthetic). Such structures are the a prioris constituting the model used to render the phenomenon intelligible or to *determine* the object of understanding. Hence, Unger and Smolin argue that even the laws of nature must be held to change in response to how they are described, reversing the scientific commitment from a realism of the static multiverse or the fixed superposition—the "block-universe"—to a realism of open process. It follows that, in their view, mathematical models are *constitutive,* at least partially, of our encounters with the universe. Against the Platonist interpretation of mathematics, the scientific practice itself is here held to conjure into existence the mathematical object, rather than *discover* it ready-made. For instance, in Smolin's view, transfinite numbers were "evoked into existence by Cantor's invention of the diagonal argument."[24]

It is telling that Smolin is a long-time friend of Carlo Rovelli, with whom he was instrumental in the development of the theory of *loop quantum gravity.* For, Rovelli is well-known as a defender of quantum relativity and for insisting that time is an illusion caused by the observer's ignorance of the microscopic states of the world. He argues that time should be understood much like the phenomenon of heat. Though we usually think of thermodynamics as the study of how heat dissipates, it can also be thought of as the *study of macroscopic objects.* The coarse-grained object is made up of many little parts, so many parts that it would be impractical to account for them all at once.[25] The macroscopic entity these little parts compose is much easier to keep track of, as the only practical way of considering their microscopic behaviors is "statistically." Thus, thermodynamics deals with a probabilistic understanding of the object in its composition. So, Rovelli argues, we observe heat on the level of the macroscopic object, but we know that heat is caused by the micromovements of the object's constituents, the atoms, which are not hot in and of themselves. The heat is an emergent molar effect of their molecular dynamics. Therefore, he argues that time too happens only on the level of the macroscopic whole, where many covariant variables are concerned, rather than on the level of the individual parts.[26]

But, as Smolin might argue, Rovelli's denial of time will always struggle with the question of the reality of his explanations. For, when one tries to define how the temporal macroscopic phenomenon emerges from the atemporal microscopic substrate, one is forced to supply a "pretemporal" account of just what had to "take place" or to "happen" in order for time to appear to be the case. More precisely, on Rovelli's account, thermodynamic time is linked to the noncommutative structure of quantum physics, which implies that "first operation A, then operation B" is not the same as "first operation B, then operation A." In noncommutative algebra, xy does not equal yx, as it would in the commutative algebra we most commonly interact with in "real world" situations. It was Werner Heisenberg who proposed to think of quantum physics as noncommutative in order to avoid the infinite quantities that arose in field theories of the time. After the further development of noncommutative geometry by Alain Connes and others in the 1980s, the notion gained traction in several aspects of physics, from string theory to general relativity. But of course, the *sequence* of operations on which the distinction between commutativity and noncommutativity rests already assumes a kind of temporal logic in which one operation comes "before" the other. It seems that, in other words, in order to claim that time emerged from nontime, we need to smuggle in an implicit temporal logic of asymmetrical process, procedure, or *operation* into what we deem atemporal. Thus, even by positing noncommutativity, we are merely replacing an asymmetry of time with a geometric asymmetry, which amounts to the same thing. For, what is time other than an asymmetrical, irreversible dimension of space? One is, in other words, still trafficking in a temporal logic. Indeed, it is as though a certain *time without becoming*, a *logical time*, or a *static genesis* is presupposed as the condition of the emergence of lived time.

This again suggests that we cannot disentangle the observer from the stratification of nature and the unpredictability of its events. But this is not a reason to interpret emergent and chaotic phenomena as illusions. We say that, if we could see the microstates and track their microeffects, there would be no apparent emergent behaviors or nonlinear processes. Every effect would have a clear and distinct cause; there would be no whole or generality, only particulars. But, according to the paradox of Laplace's demon, this hypothetical situation would simply not be a world as a matter of definition. Science can be only the science of a world, and only of *this* world that happens to be the case, only of compossibility, rather than possibility. The world and any science of it thus turn on the irreducibility of observation, which is situated, oriented, and uncertain. A universe that contains no observers is simply not (yet) a world. It is a category error to posit a world without observers: that would amount to a model of objectivity or a world model that can be isotropic and complete, a full set of well-defined causes and effects. But our models are just models, not the world itself. As Deleuze argues in reference to the responses Descartes made to Arnaud, the complete object is only "part of the object," its ideal part; on the level of the quasi whole, however, the object remains incomplete.[27] And so accounting for the privacy of subjective perception always leads us to the positing of a *quasi-causal* determination as inseparable from the conditions of aesthesis, just as attempting to explain temporal asymmetry requires us to conceive of a logical time, a static genesis, or a time without becoming.

The following dilemma might now be put forth: are the realists those who believe that there is, ideally, a model corresponding accurately to the world as it is, out there, observer-independently? Or are the realists rather not the ones who know that the model is always a correlate of observation and who thus posit a world as never fully, completely, and faithfully captured by any totalizing model? The belief that any model has captured the real in its totality necessarily depends on overlooking some of the details. Rigorously speaking, the doubt that potentially infects knowledge and action, and which threatens us with solipsistic sentiments, is also always lurking behind the clarity of the scientific model of objectivity. As Hume so brilliantly observed, we have only a collection of independent observations: we find patterns between these, but no immanent observation ensures that these patterns have any bearing on the way things really are "out there," that their stable relationships extend beyond the empirical present. Believing that our inferences have any purchase on the state of affairs amounts to little more than wishful thinking. If we are honest with ourselves, and *Pyrrhonistically* rigorous, we must admit that this world, insofar as a world is composed of many compossible perspectives upon itself, necessarily contains regions that are clear and regions that are obscure, regions that are distinct and regions that are confused from point of view to point of view. And it is according to this strictly more rigorous kind of realism—the realism of incompletion—that science and cognition can be said to take place.

To produce a world, nature "coarse-grains" itself. It produces macroscopic, coarse-grained, rounded-off, averaged-out, and approximated perspectives on itself. Indeed, this is our own intimate predicament: as products of nature, human observers face an uncertain chaotic world, within which we carve out for ourselves a metastable milieu, where predictions work with relative dependability. Insofar as it is observed and observable, nature continually separates itself into expression and content, form and matter, figure and ground. The phenomenon, the behavior, always emerges from a substrate. Worldly objectivity, the realm of *contingent truths,* always separates itself into micro and macro. The world is stratified between levels of granularity, a fact that is inseparable from the nonlinearity of trajectories in time on the macroscopic scale. Behaviors between levels of emergence resist neatly mapping onto each other. There is an *incomplete transcoding* between regimes of intelligibility, and therefore between any two regimes of periodicity, regularity, or provisional stability, there is a boundary where predictions fail, for the discernments they depend on are dominated by indistinction.

Symmetry Breaking and Experience

I have already suggested that the aesthesic integration does not concern only the concrescence or *coming-together* of the disparate. Though the general relation of togetherness is indeed necessary, it seems that it is not sufficient. There would be no event if there were not a break from the past, an expressed difference from that which preceded it. Merely being linked or associated, being related-to or together-with other elements, is not sufficient for an event to be an event, because it must also obtain the general character of a spontaneous catastrophe, where the causal isotropy is broken, resulting in an anisotropy, *an improbability* that seems to retroactively instantiate its own relatively less probable background. Aesthesis thus also concerns *symmetry breaking*: without the broken symmetry of some prior unity, superposition, or random distribution, there can be no integration to speak of. This corresponds to a generalization of the prosthesic supersession we encountered in chapter 3 that doubles the aesthesic intercession in the domain of evolution and technogenesis. But, whereas intercession and supersession can be said to occur *in time,* there is a sense in which the entanglement and symmetry breaking we observe in cosmological and quantum considerations concerns what we have referred to as *logical time* or *static genesis.* The absolute past, the first cause, will always resemble what Deleuze referred to as the Aion's fracturing of Chronos. The event always occurs *as* a break in symmetry.

These two complementary aspects of the *quasi-causal* mechanism, I believe, echo what Karen Barad refers to as "cutting together-apart,"[28] for it is as though the togetherness of that which is integrated cannot be dissociated from the break in symmetry it accompanies. "If the apparatus is changed, there is a corresponding change in the agential cut."[29] For, from the point of view of the atemporal catastrophe that divides the world into indeterminate and determinate, there can be no pregiven apparatus: it too is the product of this atemporal static genesis. Indeed, we know from Wheeler's *delayed-choice experiment* that, in some sense, the past does not exist until the experimental event retroactively settles it each time we make a measurement. A variation of this experiment known as the delayed-choice *quantum eraser* shows that there is even a sense in which we can *rewrite the past* after it has been settled by a prior act of measurement, by selectively *forgetting,* or *erasing,* the result of the measurement. Itself a product of the cutting together-apart, the apparatus, like the relativized Kantian a priori, is not the cause, but the *quasi-cause* of the event and its dual aspect of integration and symmetry breaking. Starting from any present, locate any arbitrary point on the *mathium,* what we might call the transcendental complex of possible transformations, akin to the fixed and timeless reality of Augustine. Observer-independently or apparatus-independently, from this point onward, interconnecting events ramify in all directions indiscriminately according to causality. Strictly speaking, then, there is no apparatus "before" the cut that immediately divides the field into here and there, then and now, this and that. To integrate is also to exclude, to select, and thus to break the superposed symmetry of these possible paths. But this cut then propagates. Once it has entered the causal realm, or once the Cartesian "I think" and "I am" have been codetermined by the temporal asymmetrization, then all of compossibility, all of causality, is instantaneously reorganized and reactualized. Since an experience must be materially inscribed in the organism's functional states and patterns of behavior, the event retroactively can be nothing other than a break in the previous causal topology or its world model, its Quinean total science. Events experienced less intensely, we must speculate, are retroactive changes in the microstructure of this matrix of causal lineages. Intensely lived events, for their part, are to be thought of as more substantial reorganizations of the causal topology that constitutes the settled past implicit in each asymmetrical present. This paints a picture of *symmetry breaking qua experience.*

But then this begs a question: Could the pervasive assumption of the "unity of origin" not be explained by this mechanism of the event? Just as, in the Copenhagen interpretation, the act of measuring "collapses" the wave function, deciding the value of the measurement by selecting between the quantum object's superposed states, so also here, each event of experience, even in its most ambiguous instances, proceeds as an unfolding of a singularity. There is a real sense in which each event of cognition is an echo of the spontaneous breaking of the "original" symmetry, but incongruously also a sense in which there is no true "original" break, since strictly speaking, experience starts *after the break,* which is why we can always be in doubt about the observer-independent reality of the prebreak unity. Consider again the paradox of the omniscient Laplacean demon: if one could simultaneously see all events in time and their deterministic interdependencies, nothing would happen: there would be nothing to which anything *could* happen and nothing that could happen to it, no observer and no observation. Thus, cognition is inherently *postsymmetry.* This is, in fact, quite congruent with Derrida's observation that supplementation blurs its origin, that the trace is self-effacing, as well as with Deleuze's campaign against the subordination of difference to identity, which can be seen as a call to resist the temptation to believe that, "before" the actualization, all was indeterminate.

In another sense, though, cognition is also always *presymmetry;* we observe, for instance, the universe deterministically evolving toward the hypothetical symmetry of *heat death,* where all matter will have broken down and all energy have been dissipated, leaving us with a uniform distribution indistinguishable from the singularity supposed at the beginning of time. In this way, aesthesis is not just postsymmetrical but also presymmetrical. More precisely, it is *intersymmetrical or transsymmetrical.* Even though cognition is constantly producing provisional symmetries and identifying coarse invariances, aesthesia, in its intrinsic asymmetry and incompletion resulting from the entanglements of organism and world, is drawn out *between* symmetries, distributed across them and irreducible to either of them.

So, if aesthesis is always, by definition, *transsymmetrical,* inherently in between symmetries, then what exactly is the status of the constantly reinstantiated unity, randomness, hyperchaos, or superposition of the origin? Deleuze's *transcendental empiricism* tackled problems very close to this: he can be said to have challenged the common assumption that the "great outdoors" is indeterminate, insisting instead that the virtual is "absolutely determined." This argument was necessary, he judged, to give difference a positive value, to free difference from its subordination to identity. Crucially, this meant that, instead of a genealogy originating from a unity or indeterminacy, Deleuze reversed the process. For Deleuze, the whole mechanism of causality follows from this principle: "intensity defines an objective sense for a series of irreversible states which pass, like an 'arrow of time,' from more to less differentiated, from a productive to a reduced difference, and ultimately to a cancelled difference."[30]

The arrow of time, for Deleuze, which leads from differences to their reduction and negation, corresponds directly to the process of habit formation (the first integration we looked at in chapter 3). He relates Ludwig Boltzmann's own recognition of the congruence of the second law of thermodynamics with our common-sense assumptions of time: in a system, one can generally identify the past with the improbable and the future with the probable.[31] Causal time seems to naturally go in the direction of the reduction of differences, which is the essence of the *relaxation process*: from asymmetry to symmetry. But Deleuze argues that there is a bias implicit in this process that disguises what is truly being reduced. Echoing Gilbert Simondon's notion of metastability, *the preindividuality of the process is maintained,* because, for Deleuze, *difference is never given as such.* Rather, difference is that which gives. The given is always its provisional product. Only *diversity,* he suggests, is truly reduced in the causal process from the asymmetrical to the symmetrical. Difference somehow stays intact; it never gets used up, surviving all possible transformations. We might say that he reserves the term "difference" for a somewhat paradoxically immanent *noumenon* of the *phenomenon* of diversity; it is both that by which the given is given *and* that by which the given disappears, slipping through our fingers. So, for Deleuze, this "sensible" habitual assumption that time always flows from the improbable to the probable, from the different to the same, is a misapprehension. Rather provocatively, he takes this argument so far as to call the second law a "transcendental illusion," citing an obscure interpreter of early thermodynamic theory, Leon Selme.

Taking this cue from Deleuze, we can suppose that the "magical" unity we assume of or project onto our origins, this singularity at the beginning of time, may result from our particular circumstance of being polarized within processes of symmetry breaking qua experience. Interestingly, though Simondon asserts that the primitive "magical" thought is prior to the division between subject and object, and to the field of objectivity as such, he claims nevertheless that it is populated by "key-points," places and moments that are filled with intensity or power to affect and to be affected.

> The magical universe is made up of the network of places providing access to every domain of reality: it consists of thresholds, summits, boundaries, and crossing points that are connected to one another by their singularity and their exceptional nature.[32]

Thus, Simondon is aware that the symmetry itself must be broken; the primitive preobjective, presubjective world is not a homogeneous oneness, but perhaps a previous settled actuality, the animal or prehuman world models. It is not an indeterminate transcendental unity, but rather an immanent set of preestablished symmetries that structured the causal world of individuations previous to that of hominization. For, indeed, he suggests that the primitive world is always already filled with limits that naturally have power to affect humans and/or can be more readily affected by humans. These places and moments of provisional privilege mark the causal structure of the primitive universe. The mountain top, the cave, the coast, the first moon, and the longest night, all points of key interest as geographical thresholds or temporal cusps, become "enchanted," so to speak, with inexchangeable singularity, a kind of presignifying symbolism. These key-points resonate together and form a reticulating field, a network of intensive points, limits and attractors.

It is as though we are condemned to posit an arche-trace, an ur-apparatus. This account resonates with the fact that, for cosmology, time begins not with a leap from nothing to something (from time zero), but with an origin that is *always already* anisotropic. Indeed, according to the "inflationary" model, which has recently been vindicated by the detection of gravitational waves, all structures in the universe result from the amplification of tiny fluctuations present in the inflationary beginning, when the entire universe was a microscopic quantum object. And so, when we look back in time (that is, when we look into deep space), there is a necessary *anisotropy of the background,* as is the case with the radiation from the early universe fossilized as the cosmic microwave background. We are forced to start with some initial source of variety, and so the clock cannot, as a matter of principle, be turned back to "time zero." Our mathematical extrapolations fail at such singularities, the trajectories bifurcate unpredictably, and so time zero can only be approximated. Causal time recedes, at the limit, into a logical time, a time without becoming, or a static genesis.

As Deleuze stresses, according to the *quasi-causal* mechanism of the event, this difference that gives the world is not consumed in the process.

Is this not what cosmology now also claims: that the same *quasi-causal* incursion *from nothing to something* happens everywhere around us, between and within the atoms that compose us, in the vacuum's perpetual quantum production of particle–antiparticle pairs? There is a constant production of novelty permeating the void, pure unrest and random fluctuation in the depth of existence. Can this quantum fluctuation not be equated with a static genesis, a pretemporal dynamism? It is as though the ancient theory of the *clinamen* has been vindicated as the first cause of experience and causality, the sufficient reason of cosmic emergence.

The ancient atomists asked how atoms form composite objects. Lucretius proposed a response: the concept of the *clinamen*. It was a kind of random deviation, a swerve off course that made atoms fly in random directions, rather than fall straight down like rain. The clinamen can indeed be thought of as a conceptual precursor of the now central physical phenomenon of *quantum fluctuation*. Random quantum fluctuations seem to be the physical, causal reason why all things are moving and changing, why symmetries break in the first place. It is thought that space and time, matter and energy, and locality and causality emerge from this insatiable unrest in the depths of the void as spontaneous breaks in symmetry that lead to constraints and tensions on macroscopic scales. Like little arrows, innumerable different intensities "point" at each other and redirect each other's influence, and in some places, there are inevitably some directional loops. The feedback loop leads to retroactive causations: the output is relayed to the input, and so the product is its own raw material. Depending on the surrounding conditions, some of these loops will grow (positive feedback), selectively gobbling up what allows them to persist until the favorable conditions run out. Some of these loops, however, will be conditioned to shrink (negative feedback) and converge on an arbitrarily small region of the field; and in yet other places, two of these loops will structurally couple to each other, one diverging and the other converging, finding a compromise somewhere in between (limit cycle). Instances of this third kind will lead to stable periodic behavior: oscillations, rhythms, pulses, repetitions, homeostasis, and so on. Opposing vectors neutralize each other, producing spatial symmetries or invariances. Much like *saddle points* in a vector field, they begin to constrain things around them, stabilize flows, and produce dynamic, periodic rhythms. As soon as this happens, in this mess of different influences contingently acting on each other, we are able to distinguish features in the field. In some regions, the features seem to flow unilaterally in linear propagation or diffusion. In other regions, following the lines from cause to effect seems to produce paradoxes: the cause feeds forward onto the effect; the effect curls back onto the cause.

All of this order will emerge from a very minor, indeed almost trivial, deviation from symmetry. It seems that all we need is the *clinamen* itself to account for both variation and constraint. We go from an open system of infinitely diverging (in the future) and asymptotically converging (in the past) causal chains to the emergence of *operationally closed* systems in which effects circle endlessly back into their points of bifurcation. The feedback loop is conditioned by everything around it. In neighboring regions, each action seems to cause a reaction in a causal chain that propagates and dissipates through the system, like dominos or billiard balls. But in this local region, causality loops back upon itself. Here, causality is qualitatively different from places where there is no causal recursion. Here, the chain has looped back upon itself. And we observe a distinction between two incommensurable causal regions: zones where causality is linear (the nonliving) and zones where causality is circular and teleonomic. But immediately it is as though the linear has been retroactively instituted as a ground for the circular, as Schelling realized: "Organic and inorganic nature must reciprocally explain and determine one another."[33]

As we have repeatedly seen, biology is the canonical paradigm of logically cyclical constitution. Organisms are regulated by complex systems of interlocking autocatalytic sequences, limit cycles, and dynamic equilibria. *Homeostasis is where the heart is.* The biological is founded in the principle of metastability, in maintaining the delicate and interoperating causal loops of life, and it implies, in turn, the principle of *necessity.* The emergence of such self-referential periodic behavior is the origin of necessity. From contingency emerges necessity, at the point where the field has been deterministically forced to pursue the reproduction of a certain given feature. Everything is falling into basins of attraction. Everything is determined by the field of contingent inertias converging in this or that place. But the living and the nonliving are *structurally coupled*; the circular and linear logics are forever locked in a deterministic dance. This implies that, sometimes, the shortest route to equilibrium is through a complex convoluted path involving many twists, turns, and detours through the particular exigencies of a given homeostatic feature of the landscape. The question of aesthesis can be posed nowhere but in the quasi chaos of this interference between life and nonlife, circle and line, aesthetic and anaesthetic.

As it constitutes a recognizable feature in the field, the closure of a chain of constraints upon itself grants it its own limited *perspective*. It is now causally distinct from all the linear processes not included in the loop, and as such, has landed on one side of the distinction. It now faces a *world*. The loop is now distinct from its environment, and in some minimal sense, it now "feels" its environment, for now events happen *to* it. Some events will have it grow; some events will lead to atrophy. It has become a local context of registration of difference that makes a difference. Indeed, pure difference requires the loop in order to become information, whether it be diversity, distinction, or degree, for the loop affords "sensitivity" to events in the loosest, most general sense. It is a potential sub-ject, a potential context for the registration of an event or a "datum." There could be no datum if there were not a principle of necessity, a repetition or loop, to act as a comparative context for a difference to *express* its difference. This emergence of necessity and perspective is also the emergence of partiality; asymmetry entails it by definition. This is why a world can never be totalized: objects will always be partially withdrawn; monads will be partially obscure; boxes will be black; systems will be closed. Blind spots will proliferate. The loop's constitutive horizon excludes part of the anaesthetic outside. And how a partiality and its correlative world come into being is precisely the question of aesthesis.

Aesthesis is nothing if not this transmutation of contingencies into necessities. Sentience is a break in the anaesthetic symmetry of the field, a bootstrapping of necessity out of contingency. There is now a certain imperative to persist, to pursue resilience and recalcitrance. A larval self is now *oriented* within the *clinamen* according to an independent attractor that acts as the repetition's private unit of measure or principle of selection. It is this closing off from the rest of the field of fluctuation that simultaneously affords perception, and perception is nothing but a local measure of the intensities in the chaos around the enclosed region. Repetition is selective; the causal loop selects according to its teleonomic attractor, its material resonance. The subjective necessity implied by this natural monad should not be confused with the common sense of "purpose" or "goal," which is usually thought of as a *unilateral* determination by which, from the inside to the outside, a subjectivity *intends* its object.[34] On that level, the selection of description remains a relaxation onto basins of attraction, a local dissipation of stresses according to the principle of least action. What complicates our common notions of purpose or intention is that this causal vortex is inextricably mixed with its attractor, and as such cannot be posited as distinct from its relative purpose. The system does not "know" in advance what it is attracted to, and so cannot be said to be purposive or intentional in the traditional sense. The finality, the attractor, what we might be tempted to understand as the system's intention, *polarizes* the fluxes on this region of the field, dividing them according to a relative measure. It now provisionally *feels* a world, however minimally, just enough to continually readjust its coupling with it and relax into its dynamic equilibrium. There is in fact no distinction between its action and its sensation: it does what it feels, and it feels what it does. Yet this region of the field now operates with a different set of constraints from those around it: it has bifurcated from its environment, bootstrapped itself out of the linear logic of the nonliving, and taken off on a tangent. It is now a provisional self. It is self-constraining and self-regulating, but not because it *knows* where it wants to go or intends what it wants to do. It continues to behave the way it does because it is materially constrained and conditioned to do so.

Simondon characterized the discontinuity between the self-integrating systems of biological organisms and the non-self-integrating systems of the physical domain as a kind of quantum leap: a discontinuity comparable to the quantized values of quantum mechanics.[35] When circular causalities emerge from linear ones, which is to say when the matter and form of physics are converted to the typical fractal organizations of life,[36] physical symmetries are broken and replaced with incommensurably different ones, self-referential and autonomous or holistic definitions that are irreducible to their parts. These circular invariances simultaneously constitute the causal loop's principle of retention and protention.[37] For, in its closure, in its retention, it is now a *self,* conditioned by phase boundaries, limits, and constraints, and it is now *protending a world* composed of the attractors in its state space and the geodesics leading to them. And yet the entire process from the original symmetry to the simple subject or natural monad must happen in a logical time, its whole sequence retroactively decided by the necessity of the organism's homeostasis. Like Bitbol's argument that, before the first autopoietic loop, all was superposition, Schelling's realization that the inorganic presupposes the organic implies that the genesis of aesthesia from the anaesthesic background happens atemporally, for time is born only with the constitution of the aesthesic system, and does not preexist it.

Life is structurally metastable, meaning it can react to changes in its space, changing itself in relation to the environment. When the flesh gets cold, when the body approaches an inferior thermal threshold, goose bumps rise from the skin, making the hairs stand taller to create more insulation, and hence lowering the thermal threshold. Conversely, when the human body gets too hot, sweat is released to dissipate excess heat. These are examples of *negative feedback,* by which the organism self-regulates in response to changes in its environment. The self-regulation of the organism extends its phase space, thus extending the organism's habitable environment. Yet this type of self-regulation can be thought of as an extension of the *metastability* typical of dissipative structures[38] not typically categorized as life. For instance, the cyclical stability of Bénard convection cells that emerges from a contingent break in the symmetry of a fluid is a structural compromise aiding the dissipation of heat. The convection structures that emerge are metastable, meaning they are resistant to smaller perturbations, much like a human body is resistant to slight changes in temperature (regulated by goose bumps and sweat). However they are not completely stable, for larger perturbations will lead to ruptures of the convection pattern, just as a living system is also sensitive to temperatures exceeding *its* phase space. In order to maintain the metastability of the whole, the macro level selects in favor of that which, on the micro level, composes and affords its continued existence. All living beings, all ecosystems, are nested interoperating, structurally coupled, and self-perpetuating deterministic cycles, trickling their way through a shared space of reciprocal constraints onto their attractors. Biological organisms are *societies* of nested processes of this kind: deterministic loops, importing potential and producing fixed or metastable structures, storing information in the complex states afforded by their variational degrees of freedom. Aesthesis is identical to the production and sustenance of these very structures.

Our experiences, expectations, and memories are the result of structural compromises, resolutions between all these coupled and nested, reciprocally constraining deterministic loops. We experience the world this way or that way because of their particular material and functional instantiations. Our personal past has the character it has because of them. They condition our outlook on the future. But in another way, these loops are not loops at all; they spiral in and out of each other at infinite scales and in infinitesimal fractional dimensions. They extend into each other, interfering and coalescing with each other. Whitehead's correction of the concept of the *subject* is thus particularly well-founded. Experience climbs into the implex of a *superject,* floating upon the combined vortices of these subjective swirls and repetitions. We are the concrescence of innumerable actual occasions, themselves their own concrescences of prior influences, through each causal loop's partial openness to that which is distinct from itself. It is by mutually expressing each other as a difference that the multiplicity of different experiential perspectives tend toward concrescence, asymptotically approaching totalization or unity. Experience is an irreducible emergent property of the system's many structural couplings and compromises, mutual expressions, and correlative tensions.

What about *qualia*? Here it is helpful to consider Whitehead's account of the "eternal object." Whitehead seems to insist on the absolute character of *qualia,* arguing that they do not "emerge" from such material toils. Specifically, he claims that they are *eternal objects*: "Eternal objects, such as colors, sounds, scents, geometrical characters, . . . are required for nature and are not emergent from it."[39] His account attributes a radical immanence to the "intrinsic essences"[40] of the world, by rehabilitating the absolute as that which is experienced in perfect intimacy. The problem, however, is that it also implies a fixist conception of experience: all experience is, as it were, already made, out there for all time, and all we do as subjects is iterate through the possible combinations of these unchanging elements of experience. It is true that a color, in its mode of appearing to experience, seems eternal. "It haunts time like a spirit. It comes and goes. But where it comes, it is the same color."[41] It is also true that qualities seemingly "just appear, of their own power."[42] But this does not necessarily mean that *qualia* are not in some sense emergent from natural process. The problem is that such an ontological reification of *qualia* as occurring *of their own power,* unsubordinated to the circumstances of nature and the contingencies of the event, ultimately results in a philosophy in which experience is untethered from natural process: the fundamental constituents of experience, even those felt most intimately *(qualia),* immutably subsist for all time and are unchanged by the toils of material contingency. From this point of view, Whitehead's description of the eternal object would seem to ultimately limit the purview of process, for these "intrinsic essences" take on the guise of geometrical or metaphysical constraints of spacetime itself. This is certainly at odds with the chaosmos we are beginning to resolve. To remain consistent with our speculative functionalist account, Whitehead's "eternal object" must be subjected to the same "principle of relativity" he accords the "decision" of actuality,[43] for *even "eternal" becomes a relative term* under these considerations. Somewhat ironically, something can be eternal only *relatively* to one event of concrescence, one act of prehension, one instance of the subject-superject.

There are ways, however, of reading Whitehead that allow for a convenient recuperation of his account. Brian Massumi interprets Whitehead's eternal objects as conditions of experience according to Félix Guattari's description of the way "universes of value . . . make their presence felt as though they had been always 'already there.'"[44] Quite appropriately, Massumi uses this to claim that "the 'eternity' of the eternal object is not in the wings of time." By this, he means that *qualia* should not be understood as being *in time* the way other things are, but rather as carrying within themselves their own *eternality*. They seem to operate *between* the static genesis that is the condition of time *and* causal time proper. If we take the notion of integration or exclusion, the entanglement or symmetry breaking, to be necessary conditions of aesthesis, then this makes sense: the eternal object is *that aspect of actualization that appears in the mode of the always already there.* This "in the mode of eternality" is congruent with the event's aforementioned *quasi-causal* transition between the static genesis and causal time. Thus, what appears in each experience to outlive the event's satisfaction must not outlive it in an "objective" or experience-independent sense; rather, it outlives it relatively, in its mode of appearance with regard to this specific event. Because every single event of experience is an echo of this always-absent previous break from symmetry, it hence is the expression of the difference that Deleuze claims is never consumed in actualization.

But this observation can be naturalized further. It must be combined with Simondon's account, for which the seeming eternality of *qualia* is explained by the "inherence of the intra-perceptive image." Simondon's work on perception led him to the conviction that the intraperceptive image ("attached image" or "virtual image," such as that of the "subjective contours" induced by the Kanisza triangle and other optical illusions) was neither phenomenological nor ontological, but rather a kind of middle ground between the phenomenal image and the preperceptive transcendental tendencies and tropisms that background subjective experience. In the same way, instead of ontologically reifying the mode of being of *qualia* as some fundamental characteristic of the universe, we can understand them as an effect of the mechanism of a specific subjectivity's emergence from its substrate, a friction between the implicit a priori background that logically precedes cognition and the contingent events that causally follow from it.

> The equilibrium expressed in the intra-perceptive image is that of the living with regard to a milieu, not that of the lowest energy level of the system; it [concerns] the coupling of two systems, subject and world; the intra-perceptive image is the key-point of insertion into the world of this coupling; it is symmetrical to the existence of the organism of the subject with regard to the limit that separates the subject and world.[45]

For Simondon, intraperceptive images concern the cut, the break, and the limit between the experiencing entity and the world. Subjective experience emerges the moment the elements of our organism are appropriately integrated. The moment subjectivity "lifts off" from the substrate, the break it instantiates leaves an imprint, an effect that reverberates through the subsequent process. The break contains the "seed" that will orient the tendencies that implicitly condition a life's continuous duration and maintains the provisional invariances, or relative eternalities, of experience. Throughout life, these preperceptive tendencies will interact with new perceptions and sensations, but they will never be *replaced* by new data so long as the ongoing *transcendental conditions* of experience are maintained in the wake of this break. The cut, the break, in what Deleuze would call its "Aionic" temporality of the third synthesis of time (not to be confused with *tertiary retention*), plants a contingent seed of variation, an anisotropy, into the new individuation it has conditioned into existence. This would explain why qualities *seem* immutable, unchanging, and eternal, in the sense of eternal objects, by virtue of the intrinsically invariant conditioning features of an individuation's ongoing process, its transcendental constraints. There is no way to change our relation to them, no way to think them differently or to move around them to see them from another perspective. This blue is always this blue. That red is always that red, but only insofar as it is still I who experiences it as that red, only insofar as the organism doing the cognizing must adopt a particular state from among its repertoire of states in order to perceive the corresponding color. In a sense, we take our *qualia* with us wherever we go, for our *qualia* are the imprint or the negative space of our organismic constitution, the necessarily *intrinsic* scaffolding of our sensations and cognitions, and of our construction of intelligibility.[46] *Qualia* appear with a "suchness" (the isness of that red), a *haecceity* that escapes the discursive labeling of our cognitive faculties. But this mode of appearing *as* ineffable, just like the modality of eternality, results from the particular way our experiential process emerges and declares its relative, provisional independence from the material substrate. Like birth, the emergence of subjectivity leaves its scar. The cut, the break that occurs as the perception (the superject) emerges from its substrate, always already presupposes an explanatory gap. *Qualia* are the scars left on cognition and conceptuality as it lifts off from sensation, in the same way that scientific structures and invariances are the scars left on our total science after the last paradigm shift. They are the intraperceptive textures left in the wake of this break as a new transcendental *type* of subjectivity emerges from the cooperating mixtures below and lifts off from them with entirely new capacities, new programs for constructing its *tokens* of experience. This leap, this break, leaves an axiomatic impression on the consequent superjection, a mark that potentially resonates throughout our lives, producing this illusion of the absoluteness of our internal experiences, while, in truth, they are just as revisable as the scientific invariances we trace in nature. The aspects of the event that do not change or get "used up" in the event's satisfaction are only relatively immutable; they are constraints only for the operations constructing this experience. Thus, those ineffable qualities composing the palette of our encounter with the world are neither merely subjective nor purely observer-independent. They are parts of the aesthesic compromise between the self-organizing structures and Markov blankets that we are as cognizing organisms and the relative outside. But this does not mean that even these transcendental conditions of our available *qualia* do not themselves change. Indeed, they must, even though our means to strategically operate such changes is necessarily elusive from our perspective, nested within those conditions.

Thus, such an account of the immanence of *qualia* in the real can be recuperated if we adopt the view of a stratified, distributed, mobile, and flexible view of emergent supersubjectivity. The monad, as a perspective on the world, participates in the constraints on possibility: it induces to its own extent the conditions of counterfactuality between mutually incompossible worlds. But we are compelled to speculate that, though it is participant, the monad also changes under the effects of all those other participations, under the influence of all those other monads that also retroactively modify their preconditions through their illumination of the world in various ways, since it is always already the emergent result of infinite prior contingent trials and interferences between relatively "littler" perceptions. And this happens only through the synthesis and integration of the *qualia* into the emergent relative whole that effectively is the superjection of those beneath. The Leibnizian picture is radically complicated by this, for it becomes a question of a constantly renewed monadology in which, each time a monad "takes a peek" and makes an observation, the entire realm of compossibilities is renewed in their new synthesis. We thus keep the idea that perception participates in the evolution of cosmos and is coconstitutive of objectivity, while also allowing those participant perceptions to be *quasi-causally* renewed in the universe's response to their probings and observations. This may imply that we are not condemned to the "claustrophobic" view of correlation, but rather that our transcendental types are *mobile* and that we navigate, through each break in the invariances of our settled actuality, a chaotic terrain that is truly real and independent of us.

In Search of Perceptronium

If aesthesis is of the world, inscribed in the material toils of the universe, and at the same time concerns an essential asymmetry and incompletion, then what are the minimal criteria of this aesthesic matter? What characterizes "perceptronium"? This convenient term, which highlights the material constitution of sentience, is borrowed from MIT theoretical physicist Max Tegmark, who attempts to formalize the notion of *consciousness as a phase of matter,* alongside solids, liquids, gases, plasmas, and the other more exotic states of matter, within the information dynamics of quantum cosmology. His interesting take on how experience could be inscribed into the toils of matter resonates with various aspects of our discussion of how aesthesia might emerge from a nonaesthesic background. As a staunch defender of a radical Platonic view of mathematical physics that holds the multiverse to be more than merely written in the language of mathematics, as Galileo claimed, but to actually *be* mathematics—what I have been referring to as the *mathium*—Tegmark attempts to understand how it is that conscious experience happens to take place as a specific mode or part of this monstrous mathematical multiverse. His quantum mechanical construal of consciousness as a state of matter brings together two theories: the quantum theory of *decoherence* and Giulio Tononi's theory of *consciousness as integrated information,* which is a psychological theory meant to guide the search for the "neural correlates" of consciousness.[47] Lets take a closer look at these two theories.

Early quantum theory underscored that, when we make a measurement at the microscopic level, our very observation determines the state of the system. This was typically characterized as the collapse of the wave function, where the superposed properties of the quantum object were selected through some mysterious principle, their symmetry being broken unpredictably. The spookiness in this interpretation prompted speculation about other ways of theorizing the Schrödinger equation. One of these original interpretations was Hugh Everett's. Sometimes called the "many worlds" interpretation, the Everettian interpretation implies that, in reality, the wave function never does collapse: even after we make a measurement at the microscopic level, the superimposed states of the quantum object still exist somehow but are just hidden from view; the other states have been split off into other branches of reality, independent universes corresponding to where the measurement came out otherwise or did not take place at all.

This interpretation has become key to unlocking the mysteries of decoherence. To understand decoherence, we must first discuss coherence. Coherence here refers to the properties of the wave function. From our classical point of view, since it appears as a probabilistic superposition of states, the quantum object is described succinctly as a wave-like probability distribution, describing the probabilities that the state will come out this way or that way if you do take a peek. Like all waves, then, this wave function exhibits coherence. Anyone who has worked with analog sound synthesis will intuitively grasp what this means. It means that waves can *interfere* with each other, that they can be synthesized together; their phases can be combined to amplify, modulate, or cancel each other out. By contrast, two billiard balls cannot be "modulated" together: if you crash them together, they do not form a larger lump or cancel each other out; they are individual objects and will deflect in opposite directions. Billiard balls are, in this sense, not coherent.

*De*coherence, then, is the process that leads from coherent wave-like behavior of the smeared-out superposed states of the microscopic world to the localized, object-like states we observe in the macroscopic world. In Deleuzo-Guattarian terms, it is the question of how we get from the "and-and-and" of quantum reality to the "or-or-or" of classical reality. It responds to the following question: if everything around us is in reality made of a nonlocally distributed superposition of states, smeared out across time and space, why is it that we see and interact with billiard balls and not billiard waves? If everything is actually here *and* there, this *and* that at the same time, then why does everything around us seem object-like, here *or* there, in this state *or* another, in this moment *or* that one? According to Wojciech Zurek, the way to understand how this happens is to do away with the anthropocentric idea of the observer. If a billiard ball seems to "hold together" as an object, it is because, long before we conscious observers ever come online to observe its location, the universal degrees of freedom corresponding to the billiard ball have already interacted with those corresponding to the environment that will eventually constitute us as observers. The future trajectories of the billiard ball and the environment are therefore correlated and interdependent from our perspective, because they have long been entangled accordingly. Entanglement, in other words, is not human-observer-dependent or even apparatus-dependent; it happens everywhere in nature, wherever systems interact with each other. Quantum physics is democratic in this regard: an "observer" can be any system, even a single subatomic particle. Systems "observe" each other in the same sense that our measurement devices "observe" the quantum object. And at least from our classical point of view, such entanglements with the general environment are irreversible. "While the ultimate evidence for the choice of one alternative resides in our elusive 'consciousness,' there is every indication that the choice occurs much before consciousness ever gets involved and that, once made, the choice is irrevocable."[48] We can never get the billiard ball's wave-like coherence back, but this is not because the wave function has collapsed, but rather because, according to the theory of environment-induced decoherence, the superposed states in which the particles in question did not interact and did not get entangled are statistically suppressed from our point of view.

There is a recent American science-fiction film that plays with the notion of decoherence: James Ward Byrkit's *Coherence* (2013). Unfortunately, the film gets it wrong. In the film, some mysterious astronomical event has caused a quantum disturbance that has created multiple slightly different copies of all the characters, as if multiple parallel worlds have collided. Confusion and violence ensues as the characters try to eliminate the clones of themselves that now inhabit the same hybrid world in order to regain what they deem to be "coherence." But in fact, what they are really after is decoherence. For, the coherent state corresponds to the state of superposition where the multiple worlds are confused, whereas it is decoherence, as a process, that progressively accounts for the divergence of one world from another.

From the point of view of the coherent whole (which we may understand as the multiverse, the set of all parallel worlds, the superposition of all superpositions), the wave function never collapses. The classical, macroscopic object, like the billiard ball, only seems to be a distinct object; because it is macroscopic, thus involving a great number of variables, the probability that the particles making up the ball are already entangled is extremely high. Decoherence may be the fastest process in nature; indeed, it happens as a transition from the atemporal to the temporal: a static genesis. (Notice the resonance with Rovelli's account of how time emerges from the atemporal quantum reality.) It is assumed that the reason that things look different at the quantum level, compared to the classical level, is not that smaller things necessarily look this way, but rather that, since you are looking at less stuff when peeking in at the micro scale, the probability of that stuff having already interacted with the greater environment is much lower than when you are looking at a bigger object made up of much more stuff (thus exponentially increasing the probability that its constituents are covariant with the environment). So, a macroscopic object is object-like because, the bigger something is, the higher the probability that its constituents have future trajectories correlated with those of other things in the environment. This implies that matter as we know it *is* matter because the cosmic degrees of freedom that correspond to a given object represent features corresponding to other objects: they share *mutual information.* When two particles are entangled, doing something to one of them will always affect the other correspondingly, and therefore, to know something about one particle is also to know something about the other, since they share mutual information. This insight compels Tegmark's interest in Tononi's theory of consciousness as integrated information.

Integrated information theory (IIT) is an extrinsic, information-theoretic account of conscious experience. Devised to closely mirror the *intrinsic* or phenomenological account of experience, IIT derives what it calls its "axioms of consciousness" from phenomenological observations and then attempts to reconstruct what these conditions would look like from an extrinsic perspective, in an information processing mechanism potentially observable in the world, in order to attack the famous epistemic gap between mind and matter, what David Chalmers calls the "hard problem" of consciousness. Tononi begins by listing the indubitable *axioms* of first-person experience and comes up with the following set of criteria:

Consciousness is compositional: it is always a synthesis of several things (attributes, features, forms, characteristics, *qualia*) in composition

Consciousness is informative: it tells us something meaningful about the state of the world.

Consciousness is integrated: its components cannot be subtracted from the composition without altering the holistic picture into a wholly other experience.

Consciousness is exclusive: it suppresses features of the world; it excludes certain things from the integrated or composed whole; it is always *this* experience, rather than that one.

In a second step, IIT translates these axioms into postulates, describing what we should look for when trying to extrinsically identify consciousness in systems. The postulate of composition suggests that, if experience is composed of various elements, then the material mechanisms and structures that give rise to experience will probably also be *composite mechanisms*. The postulate of *informativity,* following Gregory Bateson's famous adage, implies that "a mechanism can contribute to consciousness only if it specifies a 'difference that makes a difference' within the system."[49] In IIT, this "making of a difference" is taken to mean that the information involved must constrain the causal possibilities of the system: it must limit the possible effects that will arise from possible causes. The system must hence be characterized by a "cause-effect repertoire," in Tononi's words. The more a system configuration constrains the output effects on incoming causes, the more selective it is, and thus the more *cause-effect information* it has.

The postulate of *integration* implies that the mechanism giving rise to first-person experience must specify a cause-effect repertoire that is "*irreducible* to independent components." This is perhaps the crux of the matter, and indeed, this postulate gives its name to the theory, for it is here that we find its quantitative index of experience, referred to as Φ (*phi*), which is "assessed by partitioning the mechanism and measuring what difference this makes to its cause-effect repertoire."[50] The phenomenological argument for integration is that experience is somehow always whole: though it does seem composed of various features, *qualia,* or attributes, the experience itself is irreducible to any of these. An example Tononi gives is that the experience of a red square is irreducible to a square shape *plus* the color red; or again, the experience of the word "SONO" written in the middle of a page is irreducible to the word "SO" written at the right edge of a half page *plus* the experience of the word "NO" on the left edge of a half page.

Finally, the postulate of *exclusion* implies that, in a complex of different mechanisms, the only one that will constitute *the integrated experience* is the one that has the maximum value of integration or Φ. This implies that, in an assemblage of machines, the mechanism having a *maximally irreducible cause-effect repertoire* (MICE) suppresses all the cause-effect repertoires of the other mechanisms in the complex, those that have a lower index of integration.

Now, one of the major claims of IIT is that it constitutes a step toward solving the "hard problem" of consciousness;[51] namely, beyond the reductive description of information processing and computation, it is the hard problem of explaining the "first-person" experience that seems to be correlated with such external phenomena. IIT does seem to constitute a noticeable step beyond the Turing test, which comparatively gives us only a behavioral account of what it takes to imitate consciousness for an unsuspecting human. The canonical "philosophical zombie" could potentially pass the Turing test, making us believe it is conscious, without really being conscious at all. The Turing test apparently tells us nothing of whether some information is reserved for the perspective of the being in question. It merely shows us that certain procedural behaviors will fool humans into attributing experience to an unconscious system. IIT, however, is about explaining how a first-person perspective corresponding to what appear to be the "phenomenologically indubitable" truths can emerge from the structures and relations of the material world. It proposes a way to identify the property of first-person experience from a third-person perspective, making it possible to at least minimally quantify the amount of consciousness happening in one system versus another: Φ.

For Tegmark, the "easy problem" of consciousness comes down to a description of the hypothetical material Norman Margolus and Tomasso Toffoli termed "computronium": a model generalizing the simplest programmable substrate capable of computation or information processing. The "hard problem" of consciousness refers to the prospect of describing "perceptronium," an exotic phase of matter that "feels" the information it is processing from its own intrinsic perspective. In a sense, the goal of IIT is to formalize and model the distinction between *computronium* and *perceptronium,* to distinguish between a system that computes and *perhaps* behaves intelligently and a system that computes and has an *intrinsic* perspective on things.

Contrary to what some of the proponents of IIT have been saying about it, then, the interest of the theory is not to demonstrate that consciousness is everywhere. Christof Koch, for instance, mischaracterizes IIT as a theory of *panpsychism.*[52] But my own, as well as Tegmark's, interest in the theory stems directly from the opposite observation: it does not blindly attribute conscious mentation to all things; rather, the whole point of the theory is that it may allow us to extrinsically discriminate between something that has an internal perspective on the world and something that doesn't. Since IIT implies that all but the most trivial systems in nature will have some degree of Φ, we may refer to it more accurately as "panpsychism minus 1." In other words, it allows for what Meillassoux calls the *ancestral* realm, a material reality that is nonsentient and preexperiential, while also suggesting that the emergence of observation is not at all exceptional, but rather quite ubiquitous. Thus, it intersects one of the objectives of this book, which is to characterize aesthesia *beyond the human* and to be able to distinguish it from other things. It just so happens that, according to the theory, to get a minimal amount of first-person experience of events happening, you do not need much complexity at all. The minimally conscious system Tononi proposes is just a photodiode connected to a detector that updates the sensitivity of the diode according to its last state. It is basically a negative feedback loop with the environment, reflecting some of the insights from cybernetics and systems theory. But IIT does accept that certain systems might have zero Φ, and therefore admits that certain materials and objects, even systems exhibiting very intelligent behaviors, are not conscious. For, in IIT, there are some things for which there is no internal perspective of the world: dead matter.

It is furthermore quite significant that IIT defines consciousness as simply "experience." It does not necessarily imply *self*-awareness: consciousness, in the most general sense, is simply *aesthesic integration as such.* This reflects what I characterized as the entanglement of the "two levels" of cognition, expounded in chapter 1. The two levels are not distinguished in IIT. In its commitment to describing consciousness in materialist and functionalist terms, it does away with the distinction between sentience and sapience. It blurs the line dividing the upper and lower levels of cognition, replacing it with a spectrum of degrees in the intensity of *aesthesic integration.*

IIT does make some questionable predictions about what in the world is capable of perspectivity, and it is no surprise that some of these predictions have been met with discomfort. For example, John Searle's critical review of Koch's book on IIT is rife with sarcasm about those predictions. One such claim Searle finds difficult to swallow is that, "for these authors, there is nothing especially biological about consciousness."[53] He objects: "[For them,] such information is not confined to biological systems. You also find consciousness in, say, smartphones." Searle is clearly not ready to abandon the folk notion that consciousness must be fundamentally linked to life. A related, but more sophisticated, critique has been presented by Scott Aaronson, an MIT engineer in quantum computing.

> IIT fails . . . because it unavoidably predicts vast amounts of consciousness in physical systems that no sane person would regard as particularly "conscious" at all: indeed, systems that do nothing but apply a low-density parity-check code, or other simple transformations of their input data. Moreover, IIT predicts not merely that these systems are "slightly" conscious (which would be fine), but that they can be unboundedly more conscious than humans are.[54]

What Aaronson seems to overlook, however, is that IIT is not a theory of intelligence, but a theory of consciousness. This is important. Indeed, the theory's postulates say nothing about how intelligent a system is. Rather, it suggests that the dimension of aesthesic integration may be orthogonal to that of intelligence; these may be two independent potential behaviors of matter: *computronium* and *perceptronium.* A system might be intelligent while being completely unconscious; another system might have the most vivid experience of a certain field of potentials while having no intelligence whatsoever. Intelligence concerns effective problem solving, not subjectivity as such, and so these two potential attributes of matter are orthogonal to each other, different things matter can do. Though we humans might have intelligence and consciousness intersecting the same embodied organism, this does not mean that we will necessarily always find intelligence and consciousness in the same places when we look for them outside of ourselves, nor should we expect to find these attributes always mixed in the same proportions.

The Copernican character of IIT's displacement of what subjectivity implies is bound to be difficult to swallow. But it seems obvious that our common-sense understandings about the world need to be challenged if we are to escape the materialist limbo we find ourselves in. Our knee-jerk reactions to propositions that displace humans or (as in this case) life itself from the world stage have a long history of holding us back.

A related issue intersected by IIT's predictions is that of the "hive mind" or "universal consciousness." If integrated systems are conscious, does this mean that a society, a nation, an economy, the earth as a whole, or even the totality of all matter in the universe is conscious? Once again, within the framework of IIT, the answer is a tentative "no." The idea of a collective or global consciousness—the subject of many posthuman and new-age narratives alike—is again constrained by the theory's postulate of exclusion. For, it implies that, if ever a global or overarching consciousness were to "come online," as it were, and integrate such that it were maximally irreducible with regard to its subsystems, then all the subsystems, including our own conscious experiences, would go *offline.* Our mental integration would immediately blank out, being suppressed by the higher level of integration we are subsumed by. Luckily for us, by appealing to our bias as observers, as long as we are somewhat certain that we ourselves are (currently) conscious, we can be just as certain that a global integration at the level of the collective is not (currently) happening.

This leaves open the possibility, however, that assemblages of experiential integration occur *between the gaps* of our own waking awareness. IIT seems to allow for the momentary bootstrapping of individual substrates of perception (say, individual bodies and brains) into collective modes of awareness. This might recall the controversial notion of the "mob mentality," a collective mode of agency that is hypothetically triggered when a critical mass of collectively synchronized emotional mechanisms erupts in a crowd and seems to determine the individuals' actions from the top down. There is nothing in the theory preventing such events from coopting the experiential life of the individual in the nonexperiential gaps between individually felt events.

Be that all as it may, Tegmark's interest in IIT is the following: since decoherence stipulates that all transformations of the total energy of the universe happen in parallel, each branch instantiating a new parallel universe, then what characterizes our own branch? In Leibnizian terms, is there not something in the fact that our universe happens to contain consciousness that could help us not only identify the *possibilities* of the multiverse but also characterize the *compossibilities* of our specific universe? This reading of decoherence in terms of Leibniz is not arbitrary: decoherence conforms to the very process of transition between the possible and the compossible, articulated on the principle of nonhuman "observations." Our subjectivities and their correlated worlds are constituted by the contraction and entanglement of these *little perceptions.* Decoherence is, in effect, a contemporary *monadology*: infinite perspectives on the real codetermining each other and, in their entanglement, becoming constrained by the sieve of their compossibility. Tracing away the worlds in which their encounters happened differently, their contingent interactions describe the incompossibility of this world and that.

Correlating decoherence, cosmic inflation, and observation, Tegmark develops a sophisticated speculative account of why we find ourselves in a universe with such low entropy.[55] His argument exhibits resemblances to Leibniz's best-of-possible-worlds doctrine (which can also be seen as an anthropic argument). The crux of the argument is that, since each event of decoherence exponentially increases the integration or "mutual information" between the observed system (the given object of observation) and the environment system, the entropy (or hidden information) of the system exponentially decreases, meaning that we should not be surprised to find ourselves in a low-entropy world. In broad terms, what this scenario implies is that the cutting together-apart, the integration/exclusion, which we have identified as a basic characteristic of the aesthesic process, logically precedes locality, temporal irreversibility, and the stratification of the emergent levels of material reality.

Such reasoning leads Tegmark to explore the possibility that the particular unitary transformation of the Hamiltonian that we happen to live in is "finely tuned," as it were, to favor integrated information. Could this explain the *apparent collapse* of the wave function from our point of view? Could this explain thermodynamic time and locality? May the specific branch of the Hamiltonian we observe, which allows for a hierarchy of objects and materials, be implicit in the very fact that our causal patch has given way to conscious, experiencing observers? Is our universe special in that it has favored structures that not only compute information but also "feel" themselves processing the information? This is, of course, an example of "anthropic" reasoning in physics: it is grounded in the assumption that the fact that human observation takes place in the universe constrains the kinds of attributes we might find the observable universe to have, such as the age and temperature of the current cosmos. Tegmark's approach is to devise a series of toy universes in which different principles are adapted directly from IIT and tested.

Our investigation has offered several parallels with this scenario. In chapter 3, drawing from Simondon, we observed that aesthesis concerns a general "togetherness," an abstract reticulation or relationality that is the general form of the synthesis or integration. This aesthesic reticulation can be said to have a physical expression in quantum decoherence, which explains the progressive detachment of the classical localized object from the alocality of the quantum domain, where objects are smeared out, as it were, in space and time. Locality is indebted to an underlying entanglement of alocal quantum entities: a relation between pseudo objects precedes objecthood proper. Echoing Simondon's notion of *transduction,* in which the relation logically precedes its own terms, the entanglement precedes locality, for if anything behaves like a macroscopic object, it is because the degrees of freedom of the universe are entangled such that the object's future trajectories are appropriately correlated with the environment. The object doesn't just fall apart spontaneously, for it is held together, so to speak, by its specific way of being entangled with the other things we interact with.

But this abstract togetherness is not enough: IIT also demands the principle of exclusion. We have noted that this aesthesic "gravity" that brings the disparate together into concrescence is likewise insufficient: symmetry must be broken; there must be exclusion or suppression of aspects of the real without which there is neither event nor observer. Decoherence responds to this criterion, as it can be construed as an infinite series of symmetry breakings. Each event is at once plunged into a space without locality and a time without becoming, while participating in the classical world of this or that, here or there, then or now. Each event taken up by aesthesic intercession is doubled by a supersession. We recognize this in the decoherence/integration scenario too: the coming together and the symmetry breaking are complementary yet inseparable: they are precisely the "cutting together-apart" that Barad describes in her investigations of quantum social phenomena.[56]

The integration/exclusion concerns the fact that decoherence is a tracing out of those causal patches of the multiverse that are progressively hidden while the ones that will make up our own macrocopic world become entangled. This is, again, inseparable from the Leibnizian principle of compossiblity, which not only contracts the events compatible within this world but also simultaneously concerns the irreversible divergence between this world and others. It can also be considered the *quasi-causal* mechanism behind the material emergence of counterfactualities, of closed parallel systems of causes and effects that become strictly independent of each other. The actualization of *this* world, rather than another possible world, depends on a continual selection by the *sieve of compossibility*. Any time experience is made to conform to the strictures of the normative model, built on local invariances between the processes and gestures allowing their construction, it is always under the condition that the intrinsic compossibility of *this* world is maintained. Thus, reflecting the principle of *quasi-causality*, necessity cannot violate contingency, for the *compossibility sieve* selects necessary truths on condition that they obey the reciprocal expressions of this contingent world.

But the combination of this quantum-mechanical construal of consciousness with the view of aesthesis I have been elaborating faces certain challenges. It is worth seeing how these stumbling blocks might be dispelled or dissolved. There is a series of connected critiques that can be loosely grouped as variants of the critique of *possibilism*. The Bergsonian and Deleuzian will argue (roughly) that, since the possible is "postactual," an "image of the real," it is doubtful that its exploration can yield any insight into the true genesis of the actual. Whereas the virtual is opposed to the actual while being real, the possible is an extrapolation of the actual while being opposed to the real. Bitbol's critique of the decoherence paradigm as a "realism of the wave function" is actually in line with this general account. We know quite well that the scientific models extrapolate the a priori assumptions implicit in the experimental context in question: different experiments would yield different understandings of the universe, and so by reifying their constitution as real, we are committing science to the illusion of completion. Meillassoux, for his part, would no doubt argue that the Hamiltonian, as the "total energy" of the multiverse, is bound to mislead us because, rigorously speaking, owing to the paradoxes of set theory, the totalization of the whole is impossible. And finally, Smolin might object that the Hilbert space is an idealization and that the unitarity of quantum mechanics is an unfounded assumption, reminding us that the axiomatization of timeless laws begs to be explained on the ground of the principle of sufficient reason.

These critiques converge with the Bergsonian/Deleuzian notion that the possible is "produced after the fact, [that it is] retroactively fabricated in the image of what resembles it." For Deleuze, to understand the actual, we rather need to remember that: "Actualization breaks with resemblance as a process no less than it does with identity as a principle. Actual terms never resemble the singularities they incarnate."[57] Therefore, characterizing mind as information will be rejected on account of its observer-dependence, the formalizations used to describe the quantum world will be scorned for overlooking the unexplained axiomatizations and a prioris they carry with them, and the notion of totality as "complete" will ironically be regarded as only *part* of the real object, which is *intrinsically incomplete*.

But can this scenario nevertheless be rectified to help give insight into the nature of aesthesis beyond the human? If we do away with the structuring principles and axiomatizations of the Hilbert space and the Hamiltonian, the wave function and the state vector, could the general character of aesthesis as a decoherence and integration be adapted to these critiques? In other words, is it possible to recast the notion of aesthesia as a phase of matter without preferring any basis, without hypostatizing our settled biases? Perhaps it can be done, but not without modifying somewhat the view deployed by decoherence and IIT. We have seen that, after the realization that our axiomatizations of objectivity are always provisional to their subsequent revision, the neo-Kantian recuperation of Kant was to relativize the a priori. This is indeed what transcendental interpreters of quantum mechanics are obliged to do: Kant was wrong to assume that there were stable universal a prioris for all cognitive experience, but he was right, they say, to note that each event is conditioned by the modalities of the cognitive or experimental context, which brings with it its own implicit structuring assumptions.

However, when confronted with the critique of correlationism as a kind of solipsism, the transcendental correlationist is forced to say that what exists outside the correlation is nothing but *indeterminacy*: the act of cognition gives the object a determinacy that it did not previously have. Deleuze's argument against representationalism, and against subordinating the real to identity, should be considered with regard to this. In effect, he can be said to have attempted to step out of the symmetrical impasse between the positions held by Meillassoux and Bitbol, which we explored at the beginning of the chapter. For Deleuze, we have seen, the great outdoors is not indeterminate, but rather *absolutely determined*. His replacement of the possible with a "virtual" populated with determined singularities has the important effect of turning the representationalist account inside-out. Rather than being locked into the dialectics of identity—where we find ourselves forever faced with this sterile relation between the individual observer and the totality of the universe— crucially, for Deleuze, both the "individual" and the "totality" become asymptotically realized idealizations, projections out of the intrinsic incompletion of the boundary between virtual and actual. The events that express themselves to the observer are no longer merely random symmetry breakings from the unity. Rather, they are each time the nonrepresentational expression of that which precedes it: they do not resemble that which they express. In other words, the question of what came before the correlation is answered neither by indeterminacy nor by superposition, but by a determinate reality of singularities that is immanent to our own, yet *indistinct,* and nevertheless not subordinated to our constitution as observers of this or that type (defined by our constitutive a prioris).[58] The transcendental is no longer some otherworldly domain of indeterminacy sitting silently before the conscious observer determines it, but is already the result of infinite material trials and constraints. It has no origin, and its genesis is in a constant state of retroactive revision.

After having developed this argument in his early master works, *Difference and Repetition* and *Logic of Sense,* Deleuze can be said to have extended and begun to naturalize it in his subsequent work with Guattari. Together, they write of transcoding between milieus and strata, none of which can be mapped completely onto others, implying a *chaosmos,* rather than a cosmos. And so, the picture becomes quite different: we can no longer assume that all the chains of causality have diverged from some original unity in the past, nor that they necessarily always converge as a "determined" individual in the future. Rather, the chaosmos is not subordinated to our bias as observers to observe the past as unity, symmetry, or superposition, and it has an independent reality corresponding to the "static genesis" that, as we have seen, creeps into even the most relativistic rejection of temporal realism. Static genesis no longer operates in some mystical realm of indeterminacy, but corresponds to the real-world trials of the strange beings (organic, mineral, technical, algorithmic, or absolutely alien) that participate in the constitution of compossibility and have always already conditioned our waking experience by the time we enter the scene as observers. The symmetry-breaking event takes on a whole new guise: it is no longer just an iteration along a linear sequence of supplemental observations leading back to the origin and forward to the end of the universe. Rather, time is exploded into a multiplicity, a fractal topology of infinite nested milieus or determined singularities that, in their interference, produce a constantly renewed world of compossibilities.

This ontology extends all the way up to the artifacts, practices, languages, and psycho-social structures that humans construct and must be made to resonate with our "general organology," for our technically exteriorized environments too are formed of the intersections and couplings between the reciprocally transcoding milieus composing the world. Following the thesis according to which organology conditions temporality, technoscientific exteriorizations of cognitive processes ripple across these provisional structures and processes, propagating and resonating through the matrix, suggesting a radical temporal relativity for each milieu. In the strong sense, it suggests that there can be no *eternal object*, ideal form, or absolute origin, but only an infinite series of provisional propositions of *supplementary* forms, models, materials, and processes of ideation, each characterized at once by a dimension of memory and a dimension of amnesia. Perceptronium, therefore, occurs contingently wherever there are interfaces between circle and line, between subjective and objective necessity, no matter where it finds itself in the material hierarchy. This is inextricable from the question of why we are faced with a chaosmos rather than a cosmos: *kosmos* implies a cosmic order and symmetry, while *chaosmos* captures the current order's provisionality as a subset of the chaotic whirl it emerges from. Past, present, and future can be said to differ only in their respective orientations and engagements within the nested entanglements loosely forming the intermilieux, the pan-*Umweltic*, or the transobjective. And correspondingly, there is no unified origin: the universe does not reduce to some ultimate cause, but infinitely ramifies into different causal pasts.

EPISTEM AESTHETICS: A SCIENCE NOT OF WHAT IS, BUT OF WHAT WILL' HAVE BEEN.

CONCLUSION

Aesthesia in the Wild

WE HAVE CONSIDERED AESTHESIS as a process inseparable from the production of material subjectivities in the wild, beyond the human. The combined realization that logic cannot be reduced to intentional or qualitative simples and that inferences and judgments are always conditioned by the clear yet confused prepredicative discernments they bathe in and emerge from have simultaneously urged a rehabilitation of aesthesis as a condition of intelligibility. Aesthesia can now productively be construed as an integrated weave of constraints between acts of codetermination, measurement, observation, or entanglement, various encounters between aspects of the real in the construction of a web of *compossibilities.* Indeed it has become clear that aesthesis is somehow inseparable from this division between the possible and the compossible, the contingent catastrophe that *quasi-causes* the divergence of our world from the ones it retroactively originates in. From these contingent compossibilities derive the provisional invariants and symmetries of our models of objectivity, as well as the apparent stability of our *qualia:* they are the tokens of experience available to the transcendental types of subjectivity that emerge from the contingent evolution of the *quasi-causal* entanglements between relatively littler perceptions. Epistemaesthetics thus finds itself resituated as a study of the trials and the drama of material emergence, in this world, rather than any possible world. We might say that epistemaesthetics is the science not of *what is,* but of *what will have been.*

A strangeness, of course, has been found to lurk in the depths of these speculations, a retroactivity perhaps inseparable from our cognitive embedding within the world. For, it is clear that cognition requires material constraints. An abstract relationality between possible events is not sufficient for cognition to take place: some material substrate must accommodate the relative outside for a datum to be transformed into information and *make a difference*, for a possibility to become a compossibility, for a potential to be made actual. Something that does not happen *to* something is not an event in the strict sense, but only a possible event, indifferent from all the counterfactual events that equivalently might happen in other possible worlds. We have seen how the deep historical precursors of technogenesis and the organic origins of reason seem to collaborate with aesthesis in the mechanism of this spontaneous break from eternal silence at every level of description. But a question remains: how exactly do the different worlds retroproduced by each aesthesic integration combine in entanglement if there is nothing there to be entangled in the first place, if all entanglements take place indiscriminately giving rise to all worlds equivalently? We are faced with an ineliminable economy of the circle of self-relation and the line of explanation or entailment. And at every level of abstraction, from the lowest level of description to the most coarse-grained, there is an irreducible mechanism that gives rise to being and to that which is said of being, to the world and that which is perceived of the world. Aesthesis is inextricable from this mechanism; indeed, it *is* this very process: an intercession of differences that is the condition of their supersession, the engine behind the coarse-graining of nature as it bootstraps itself into a world.

The tension between philosophers and scientists with regard to the observer-independent existence of time, chaos, and emergent phenomena is as old as the incongruity between Heraclitean and Parmenidean views. Does time really exist? Does randomness really exist? Do any properties emerge on the level of the provisional whole? Or are these mere illusions, cast upon us by the fallibility of our senses and the limitation of our knowledge of the constituents? Of course, if the mathematical description of causality is objectively valid, then time is reduced to mere illusion, an effect of deterministic biases. The various issues we have encountered, from nontotalizability as a matter of principle to the contingent nature of the a priori and the axiomatized physical or mathematical law, should have by now suggested that, against the doctrine of reductionist materialism is held up a materialist doctrine of incompletion, of ubiquitous emergence and supplementation, innumerable layers of mediation, eternal regress toward infinite lost origins perpetually deferred. We saw that the complexities of evolution actually imply this: the family tree is always just an arbitrary cross-section of the family rhizome. There is no singular origin, just as there is no singular finality. Our technoscience is constantly rewriting the past, retracing the genesis that conditions the present. And the matter we observe, interact with, and co-opt into self-production is no longer that which gives our objects their recalcitrance and stability, but what actually challenges those stabilities, the reason our explanations are always provisional and our models incomplete. Matter is a principle of unrest in the depths of the void. Matter is the inescapability of time, the unreason that conditions our reasons, that challenges our most secure understanding. To give a materialist account of aesthesis is necessarily to creep into indistinction and admit the incompletion and underdetermination of all our models, all our maps. Matter, hence, also condemns us to a superstitious yet enlightened realism, each event renewing our commitment to the concerns of the world, our engagement with real states of affairs. Failing that, we would be directly opposed to our most intrinsic orientation and asymmetry within the chaosmos, a performative contradiction that, in the act, would have always already deleted us from existence, in the future perfect tense.

If there is no strict lowest level of description, no true first organism or correlation, then the distinction between fact and speculation breaks down, because the aesthesic conditioning of each event of experience infects that which will have been. There is thus an aspect of the future that is conditioned by aesthesic receptivity to events, the affective tonality, the attitude with which subjectivity retroactively integrates the occurrence. The noted effects of observation bias are formidable instances of this very conditioning of causality by the logically prior aesthesic integration. Observation bias suggests a constraint on our faith in reality; the very fact that we are here experiencing this world, making these observations and not others, limits and distorts what we might expect to find in the universe, and indeed participates probabilistically with the ongoing production of the cosmos *as a horizon of expectation.* The superstitious leap of faith, the speculative swerve that has the organism contemplate this rather than that, to ask the universe to reveal this aspect rather than that, seems to have inaugurated a universe that holds to nothing outside a purely contingent aesthesic decision of indication and distinction. Today's cosmological conundrum of how and why the universe seems "finely tuned" in just the right way for we humans to have evolved into it echoes Leibniz's "best of possible worlds" doctrine: though it *could* have been otherwise, since other worlds are possible, the *sieve of compossibility* selects a world in which *this* experience, *this* observation, is produced. The asymptote of aesthesis hence obeys a single criterion: a convergence of potentials upon *this* provisional compossible world. Contingency retroactively becomes the precondition of necessity, in the same movement that produces the integration and exclusivity of conscious perception. This world is severed from that world in the same movement that produces the continuity of experience. In each event, there is both a divergence and a convergence: an agential cut that irreversibly severs the ties between possible worlds and a reticulating concrescence between potentials that produces aesthesia as the forever-retreating asymptote of the process. The asymptotic convergence of actualized potentials in aesthesia results from the event's severing of this world from that one, each time reestablishing and reaffirming the distinction of compossible and possible.

To what strange mechanism, then, do we owe the most improbable winning streak of science? To this question, the process of aesthesis, thus described, offers a compelling response. It is the *quasi-causal* mechanism that, in the future-perfect tense, always echoes the mantra of facticity: *it will have been thus.* But this does not mean that an inexorable zero-sum game is played between mind and matter. We will never step out completely into the great outdoors, for that would imply an annihilation infinitely more serious than death, retroactively reconditioning our universe such that nothing will ever have been thus: we will never have been here; we will never have known anything or even aspired to know; there will never have been any world to discuss and debate over, no matters of concern, no problems or facts of the matter. But, though we may never escape our perspectivity, asymmetrical, oriented, and biased, this is by no means a reason to abandon the patient task of understanding, and thereby of challenging our current biases and static relations to the world. In fact, we are condemned to keep trying to skew and distort our limited perspectivity, to integrate ever more orientations and perspectives into our total science, and attempt to discover the transcendental symmetries that will allow us to resynthesize our world into an ever more hybrid form. An aesthetic-ethical imperative thus weighs on us: we must pursue the expansion of our phase space; we must avoid stupidity; we must try harder and fail again, but fail ever better. In a word, we are condemned to speculation, to *aesthesis*, the transition from indistinct to distinct.

And this, as we have seen, concerns the deeper matter of distinction between art and other technologies. Technogenesis, in its exteriorizing deferral of cognitive experience and intentionality, follows and extends the superstitious constraint on faith that all practical reasoning depends on: a rigorously *unfounded* belief that the stabilities we observe offer some kind of purchase on reality, a hope that they are more than just random flukes, occurrences untethered to any common stable basis. Evolution condemns the organism to realism and favors the creature that believes in the stability and invariance of the necessities regimenting the real. But the artistic act marshals this same mysterious superstitious element in evolutionary and technogenetic reason, turning it against itself, revealing the paradox at its heart. The artwork does not merely show us the indistinct; by enframing the act of enframing, by revealing the contingency of this act, the artwork can also be said to challenge the closure of the correlation between subject and object. It inserts itself into the flow, causing a disturbance, diffracting subjectivity into its contingent variations, and collapsing, if only for a moment, the objectivity aesthesis inevitably instantiates. In the realm of human affairs, art echoes the same principle of variation that characterizes matter as such: incompletion, pure unrest, contingency without dependence, intrinsically random mutations, and their pseudorandom recombinatorial echoes. This is why, if *technē* proper can be said to be the exteriorization of a practical form of reasoning, art is the exteriorization of a *pure reason*, speculation in its most rigorous form, a refusal to submit to the utilitarian impulse, the given model of objectivity, or the facile judgment of the state of affairs. Whitehead was right to warn us: "A self-satisfied rationalism is in effect a form of anti-rationalism."[1] Art can be construed precisely on the basis of such a logic: it is the technical exteriorization of a Pyrrhonistic *epoché*, a radical suspension of judgment, a refusal to "arbitrarily halt" the infinite regress of conditionals, expressing a provisional freedom from the constraints that otherwise condemn us to a superstitious trust in the completion of our models, in effect challenging our commitment to a static transcendental subjectivity and its immutable limits.

The Anthropocene's dilemma between vitalism and eliminativism must be reassessed in light of these observations. Aesthesis is neither intelligence, nor reason, nor agency, though it inevitably conditions their provisional axiomatizations. Intelligence concerns an extension of the biological's drive to reason and predict changes in its environment, its capacity to solve practical problems. But aesthesis is orthogonal to this evolutionarily forwarded practical reason, which concerns a unilateral assessment of the relative outside, a superstitious belief in the completion of the symmetries and orderings that compose objectivity. It has been suggested that some regions of the material world may be conscious while lacking any significant form of intelligence. Some other regions may be superintelligent while achieving no significant awareness of the world ("zombies"). Thinking aesthesis in the wild requires that we do away with this anthropocentric bias for assuming that such characteristics should always be found in the same proportions everywhere we look. Some systems perhaps invest the prosthetic much more than they invest the aesthetic, while others are integrated so profoundly that their existence corresponds to the contemplation of but one experience for all of eternity, or until the conditions run out.

It all depends on how matter is provisionally organized from place to place. Over here, the constraints and flexibility between strata allow for complex mnemomic dynamics, permitting a system to extend its phase space, deterministically increasing its causal wiggle room. Over there, the constraints are such that all but a tiny sliver of the universe is eclipsed and one event is observed for all of time. In yet other locations, there is no aesthesia at all, no integration, no exclusion. Panpsychist theory fails to account for this. If no integration takes place, if no economy of inclusion and exclusion is occurring in this parcel of the cosmos, if the matter simply isn't organized in such a way as to accommodate the retention of the relative outside, then, strictly speaking, there is no aesthesia here. This does not mean that the site in question might not simultaneously serve as a material taking part in a process of aesthesis, asymptotically culminating in one or many integrations. For, such a functionalist account presupposes the substrate independence of aesthesia: as in the ship of Theseus, the planks building up the sentient whole may be replaced with others and need not themselves be harmonic molecular echoes, all the way down, of the experiences taking place on the level of the molar entity. Though the boundaries between aesthesic processes are mobile and far-reaching, one cannot say that *all* regions, and all *levels of description* or *abstraction* are themselves aesthesic integrations.

And yet the world nevertheless appears to be teeming with incommensurable aesthesic integrations. Though the panpsychist is wrong to assume that *every* parcel of reality is sentient, the assumption is right that sentience is not such a rare thing in the material world. The pebble itself is not sentient on the level of description from which we identify it as a small rocky object, but the pile of pebbles, in its self-organized criticality, may exhibit the limit case of the primary integration of what we humans take to be the temporal dimension, on the level of its provisional whole, poised as it is on the evental cusp of the avalanche.

Scale and level of description mean everything. Aesthesia, in effect, cannot happen on a level deemed "basic": the economy of distinction and indistinction on which it rests presupposes a difference between levels of abstraction. As we have seen, a basic level is always provisional: cognition's default of origin, the logic of supplementarity, and the uncertainty in quantum physics imply that the cosmos is bottomless, groundless, strictly speaking. But any level *provisionally* deemed basic cannot itself be aesthesic because aesthesis presupposes the difference between levels of description and granularity. In other words, cognition is inseparable from an economy of distinction and indistinction. This constitutes a block on the reductionist hypothesis, though we can agree that, once the micro level's constituents are in place, so are the phenomena on the macro level; there is an epistemological limitation on reducibility having to do with the very fact that the flattening of the world onto a single stratum also implies the dissolution of cognition itself. It corresponds to why, outside the "correlation" of thought and being, we have trouble imagining anything but radical contingency or superposition. And in quantum physics, it indeed becomes unclear whether this constraint is merely epistemological or truly ontological. For, it seems a world must coarse-grain itself in the process of aesthesis. A world *is* a decoherence of the supposed consistency and superposition of the retroprojected origin, it passes from the and-and-and to the or-or-or.

Furthermore, the transition is logical rather than temporal, *quasi-causal* rather than causal. This suggests that certain aesthesic integrations may not even evolve temporally in what we biological beings take to be the temporal dimension. The recent discovery of "space-time crystals" could be held in support of the idea that an aesthesic material system could be constructed in such a way as to be stable in our time dimension but asymmetrically oriented (irreversible) within one or more of our symmetrical spatial dimensions: a transversally sentient being inhabiting a world deployed in the interstices of our own. What flows in one direction for us may run in the reverse for other beings; what is stable and reversible in our world may be oriented and irreversible in theirs, the source of events that we cannot currently begin to fathom.

The world is, as Leibniz observed, nowhere but in its various perspectives on itself. These alien participant percepts, these transversal organisms, are also reciprocally presupposed by our own aesthesic integrations within the web of compossibilities. What is to be made or said of them? How should we include them or consider them? It is in this question that we find a convergence between two other figures of the Anthropo-cene, between the promethean and the luddite, between the champion of enlightenment and the deep or "dark" ecologist, between those who want the human mind to pursue its bifurcation from nature and those who wish to return to earlier, humbler modes of being and pay heed to the concerns of the alien forms of subjectivity that abound. For, it just so happens that the only way to welcome these alien matters of concern into our political and ethical purview is, somewhat ironically, to pursue enlightenment and what to the ecologist inexorably appears as a mistake, a *fall*, a fateful expulsion from Eden. Nonhuman aesthetics, or epistem-aesthetics, indeed suggests that the two opposing perspectives actually converge: the only path to ecology is to continually challenge our static form of human transcendental subjectivity, to distort, to experiment, to speculate, and to further integrate what today seem to be incompatible, incommensurable forms of aesthesic integration into our purview.

What is crucial to keep in mind is that the "explanatory gap" between internal experience and objectivity is redoubled within the real itself. There will be no ultimate causal explanation for causality: we can only extend its symmetries and invariances further and further, driven by the question of why it will have been thus, perpetually appending new demands for prior conditions. The answer can come only in the form of a further extrapolation of causal lineages into the ever-deeper past. But, far from implying a claustrophobic or solipsistic view of the correlation, this activity itself implies a constant modification of our paradigms, of our transcendental types, hybridizing subjectivity with alien forms, creating new programs of thought and new experiential constructions, previously unthinkable percepts, affects, and concepts. For, in a sense, if we are con-demned to an *incomplete naturalization* of experience, it is because we are suspended between *two* different potential performative contradic-tions: once in the naïve realist's attempt to break free of the transcenden-tal conditions that always precede cognition, and again in the solipsist's inference that nothing at all is the case (in the latter formulation, Meillas-soux can be said to converge with Hilary Putnam's argument for why we are not "brains in a vat"). In other words, we are suspended between two untenable alternatives: naïve realism and naïve correlationism. Hence, we are condemned to a middle way: to take experience seriously (and not "explain it away"), and therefore *try* to naturalize it, while also admitting within nature itself an essential incompletion, a hyperchaotic element in the depths of the noumenal real. The point is that this operation conforms to a persistent quest to expand or distort the current limits of subjectivity and cognition, to hybridize human life with alien forms of cognition, to integrate the strangest of experiential types into our own, thereby welcoming alien matters of concern into an expanding political and ethical purview, inventing new paradigms of thought, new worlds and their participant perceptions.

If the static, finite, and ignorant form of human subjectivity is to be overcome, and with it the notion of a fixed and final set of transcendental constraints on cognition, a shift is required from the mystical exaltation of origins and finalities to the labor of describing the structures and strata that populate the transymmetrical in-between. This requires a praxis of aesthesic experimentalism and speculation that proposes new syntheses, new expressions of operative indistinction, from which new mobile correlations can lift off, from which new superjectivities can emerge to pursue the task indefinitely. Individuation presupposes transindividuation: hybridization, synthesis, and transduction. But it cannot happen on its own. It is a labor, a praxis, a discipline. We cannot sidestep the challenge: there is no easy way out, no shortcut or magic trick that can, in the blink of an eye, give us peace. We must assume our failures, revise our commitments, and always return to the interminable task, the infinite series of trials and errors that has us perpetually challenge our static subjectivities, and in so doing, continually adopt new alien matters of concern as our own. Enlightenment and ecology converge in the task of hybridizing subjectivity.

NOTES

Introduction

1. Thomas Nagel, "What Is It Like to Be a Bat?," *The Philosophical Review* 83, no. 4 (1974): 435–50.
2. Paul M. Churchland, "Consciousness and the Introspection of 'Qualitative Simples,'" in *Consciousness Inside and Out: Phenomenology, Neuroscience, and the Nature of Experience*, ed. Richard Brown (Dordrecht: Springer Netherlands, 2014), 35–56.
3. Gabriel Catren, "A Plea for Narcissus," in *The Legacy of Kant in Sellars and Meillassoux: Analytic and Continental Kantianism*, ed. Fabio Gironi (New York: Routledge, 2017), 163–64
4. Max Tegmark, "Consciousness as a State of Matter," *Chaos, Solitons and Fractals* 76 (2015): 238–70 (arxiv.org/pdf/1401.1219.pdf).

1. Chaos and Cognition

1. Gottfried Wilhelm Leibniz, *Leibniz: Philosophical Essays,* trans. Roger Ariew and Daniel Garber (Indianapolis, Ind.: Hackett, 1989), 24.
2. Leibniz, *Philosophical Essays,* 29.
3. Leibniz, *Philosophical Essays,* 24.
4. Leibniz, *Philosophical Essays,* 24.
5. Leibniz, *Philosophical Essays,* 24.
6. Nicholas Jolley, *Leibniz and Locke: A Study of the New Essays on Human Understanding* (Oxford and New York: Clarendon and Oxford University Press, 1984), 184.
7. Stephen M. Puryear, "Was Leibniz Confused about Confusion?" ed. Glenn A. Hartz, *Leibniz Society Review* 15 (2005): 95–124.
8. Leibniz, *Philosophical Essays,* 28.
9. Leibniz, *Philosophical Essays,* 96.

10. Gilles Deleuze, *Difference and Repetition* (New York: Columbia University Press, 1994), 280.

11. Immanuel Kant, *Critique of Pure Reason*, trans. Werner S. Pluhar, ed. James W. Ellington, unified ed. (Indianapolis, Ind.: Hackett, 1996), 326.

12. Kant, *Critique of Pure Reason*, 330.

13. Kant, *Critique of Pure Reason*, 74 (note to B36).

14. Thomas Metzinger, *The Ego Tunnel: The Science of the Mind and the Myth of the Self* (New York: Basic, 2010), 50.

15. Metzinger, *The Ego Tunnel*, 50.

16. Gilbert Simondon makes the link between the ineffability problem and the question of intuition and relates it to the "just noticeable difference" as it is described in Victor Egger's *La parole interieure* (*Imagination et invention*, [Chatou: Éditions de la Transparence, 2008], 79–80 and 104).

17. Kant, *Critique of Pure Reason*, 326.

18. Mark van Atten, "Kant and Real Numbers," in *Epistemology versus Ontology: Essays on the Philosophy and Foundations of Mathematics in Honour of Per Martin-Löf*, ed. Peter Dybjer, Sten Lindström, Erik Palmgren, and Göran Sundholm, Logic, Epistemology, and the Unity of Science 27 (Dordrecht: Springer, 2012), 3–24.

19. Plato, *Republic* 10.605b–c, trans. Benjamin Jowett (New York: Vintage, 1991).

20. Ralph Cudworth, quoted in William James, *The Principles of Psychology*, vol. 2 (New York: Dover, 1890), 11.

21. Georg Wilhelm Friedrich Hegel, *The Encyclopaedia Logic*, part I of the *Encyclopaedia of Philosophical Sciences, with the Zusätze* , trans. T. F. Geraets, W. A. Suchting, and H. S. Harris, with introduction and notes (Indianapolis, Ind.: Hackett, 1991), 85.

22. Bernard Teissier, "Géométrie et cognition: l'exemple du continu," November 2, 2007, webusers.imj-prg.fr/~bernard.teissier/documents/Cerisy06final5-1.pdf, 15 pp., at 5 (my translation).

23. Teissier, "Géométrie et cognition," 11 (my translation).

24. Gilles Deleuze, *The Logic of Sense*, trans. Mark Lester and Charles Stivale, ed. Constantin V. Boundas (London: Athlone, 1990), 186.

25. Salomon Maimon, *Essay on Transcendental Philosophy* (London and New York: Continuum, 2010), 93

26. David Hume, *An Enquiry Concerning Human Understanding*, Philosophical Essays Concerning Human Understanding (Chicago: Open Court, 1921), 85.

27. René Thom, *Semio Physics: A Sketch*, The Advanced Book Program (Redwood City, Calif.: Addison-Wesley, 1990), 7.

28. Teissier, "Géométrie et cognition."

29. Gottfried Wilhelm Leibniz and Samuel Clarke, *The Leibniz-Clarke Correspondence: Together with Extracts from Newton's Principia and Opticks*, ed. H. G. Alexander (Manchester: Manchester University Press, 1998; repr.), 25–26.

30. As quoted in Michel Bitbol, Pierre Kerszberg, and Jean Petitot, "Introduction," in *Constituting Objectivity: Transcendental Perspectives on Modern Physics*, ed.

Bitbol, Kerszberg, and Petitot, The Western Ontario Series in Philosophy of Science 74 (Dordrecht: Springer, 2009), 12.

31. Albert Einstein, "Geometrie und Erfahrung," in *Geometrie und Erfahrung* (Berlin and Heidelberg: Springer, 1921), 21.

32. Paul Patton, "Redescriptive Philosophy: Deleuze and Rorty," in *Deleuze and Pragmatism,* ed. Sean Bowden, Routledge Studies in Contemporary Philosophy 61 (New York: Routledge, 2015), 148

33. Hans Reichenbach, *The Theory of Relativity and a Priori Knowledge,* trans. and ed. Maria Reichenbach (Berkeley: University of California Press, 1965), 105.

34. Quoted in Bitbol, Kerszberg, and Petitot, "Introduction," 13.

35. Reichenbach, *Theory of Relativity,* 104.

36. Maimon, *Essay on Transcendental Philosophy,* 21.

37. Lewis Carroll, "What the Tortoise Said to Achilles," *Mind* 4, no. 14 (1895): 278–80.

38. Deleuze, *Logic of Sense,* 16.

39. Deleuze, *Logic of Sense,* 18.

40. Deleuze, *Logic of Sense,* 19.

41. Teissier, "Géométrie et Cognition."

42. Francis Bailly and Giuseppe Longo, *Mathematics and the Natural Sciences: The Physical Singularity of Life*, Advances in Computer Science and Engineering Texts 7 (London and Hackensack, N.J.: Imperial College Press, 2011), 89.

43. W. V. Quine, "Main Trends in Recent Philosophy: Two Dogmas of Empiricism," *The Philosophical Review* 60, no. 1 (January 1951): 39–40.

44. Wilfrid Sellars, *Science, Perception, and Reality* (Atascadero, Calif: Ridgeview, 1991), 127–96.

45. Martin Heidegger, *Being and Time,* trans. John Macquarrie and Edward Robinson (New York: HarperSanFrancisco, 2002).

46. Jacques Derrida, *Of Grammatology,* trans. Gayatri Chakravorty Spivak, corrected ed. (Baltimore, Md.: Johns Hopkins University Press, 1998).

47. Donald Davidson, *Subjective, Intersubjective, Objective* (Oxford and New York: Clarendon and Oxford University Press, 2001), 141.

48. Deleuze, *Logic of Sense,* 21.

49. René Descartes, *The Philosophical Writings of Descartes,* vol. 2, trans. John Cottingham, Robert Stoothoff, and Dugald Murdoch (Cambridge and New York: Cambridge University Press, 1984), 106.

50. Michel Serres, *Le Système de Leibniz et Ses Modèles Mathématiques: Étoiles, Schémas, Points,* 2nd ed., Epiméthée (Paris: Presses universitaires de France, 1982), 217 (my translation).

51. Gilles Deleuze, *Logique du sens* (Paris: Éditions de Minuit, 1982), 41 (my translation).

52. Edmund L. Gettier, "Is Justified True Belief Knowledge?" *Analysis* 23, no. 6 (1963):121–23.

53. E. Gombrich, *Art and Illusion: A Study in the Psychology of Pictorial Representation,* millennium ed. (Princeton, N.J.: Princeton University Press, 2000).

54. John Ruskin, *The Elements of Drawing* (Lexington, Ky.: CreateSpace, 2010).

55. James, *Principles of Psychology*, 2:9.

56. John Locke, *An Essay Concerning Human Understanding*, bk. 2, ch. 23, no. 29, in *The Works of John Locke*, 9 vols., 12th ed. (London, 1884), 1:308 (emphasis added).

57. Quine, "Two Dogmas," 39.

58. Deleuze, *Logic of Sense*, 182.

59. Steven Shaviro, "The 'Wrenching Duality' of Aesthetics: Kant, Deleuze, and the 'Theory of the Sensible,'" November 10, 2007, shaviro.com/Othertexts/SPEP.pdf.

60. Gilles Deleuze, *Kant's Critical Philosophy: The Doctrine of the Faculties*, trans. Hugh Tomlinson (London: Athlone, 1984), 128–29.

61. Deleuze, *Kant's Critical Philosophy*, 137.

62. George Brecht, *Chance Imagery* (New York: Something Else Press, 1966), 12.

63. See André Breton, "Le Message automatique," *Minotaure*, no. 3–4 (1933): 55–65; Salvador Dalí, "Interprétation paranoïaque-critique de l'image obsédante: 'L'Angelus' de Millet," *Minotaure*, no. 3–4 (1933), 65–67.

64. La Monte Young, *An Anthology of Chance Operations . . .* (New York: L. Young and J. MacLow, 1963).

65. Immanuel Kant, *Critique of Judgment*, trans. Werner S. Pluhar (Indianapolis, Ind.: Hackett Publishing, 1987),173–74 (§45).

66. "Genius," we should note, is the Latin word for the Greek *daemon* or *daimon*.

67. Krzysztof Burdzy, *Probability Is Symmetry* (Seattle: University of Washington Press, 2003).

68. Ron Eglash, *African Fractals: Modern Computing and Indigenous Design* (New Brunswick, N.J.: Rutgers University Press, 1999), 96–98.

69. Gottfried Wilhelm Leibniz, "Explanation of Binary Arithmetic," trans. Loyd Strickland, in *Die mathematische schriften von Gottfried Wilhelm Leibniz*, ed. C. I. Gerhardt (Halle: H. W. Scmhidt, 1859), 7:223–27 (leibniz-translations.com/binary.htm).

70. As Brecht noticed in *Chance Imagery*, humans are notoriously bad at choosing randomly when they want to: we are always biased.

71. Marcello Buiatti and Giuseppe Longo, "Randomness and Multilevel Interactions in Biology," *Theory in Biosciences* 132, no. 3 (2013): 139–58.

72. Stephen Wolfram, *A New Kind of Science* (Champaign, Ill.: Wolfram Media, 2002).

73. Daniel C. Dennett, "Real Patterns," *Journal of Philosophy* 88, no. 1 (1991): 27–51.

74. Luciano Floridi, "Against Digital Ontology," *Synthese* 168, no. 1 (2009): 151–78.

2. Determinism and Drift

1. Rebecca Bliege Bird and Eric Alden Smith, "Signaling Theory, Strategic Interaction, and Symbolic Capital," *Current Anthropology* 46, no. 2 (April 2005): 221–48.

2. Thorstein Veblen, *The Theory of the Leisure Class: An Economic Study of Institutions* (New York : Macmillan, 1912).

3. Étienne Souriau, *L'avenir de l'esthetique: Essai sur l'object d'une science naissante* (Paris: Alcan, 1929), 96 (my translation).

4. Jacques Rancière et al., *The Philosopher and His Poor*, trans. John Drury (Durham, N.C.: Duke University Press, 2004).

5. Denis Dutton, *The Art Instinct: Beauty, Pleasure, and Human Evolution* (New York: Bloomsbury, 2008), 51 (emphasis added).

6. William James, "What Psychical Research has Accomplished," in *The Will to Believe And Other Essays in Popular Philosophy* (New York : Longmans, 1897), 300.

7. Francisco J. Varela, Evan Thompson, and Eleanor Rosch use this term in their *The Embodied Mind: Cognitive Science and Human Experience* (Cambridge, Mass.: MIT Press, 1991) to mean adaptations that are less than optimal (satisfying) for survival but still sufficient. It goes back at least as far as Herbert Simon, who used it as a portmanteau of "satisfy" and "suffice" for a decision maker accepting a less-than-optimal solution as sufficient (see *Administrative Behavior: A Study of Decision-making Processes in Administrative Organization* [New York: Macmillan, 1947] and "Rational Choice and the Structure of the Environment," *Psychological Review* 63, no. 2 [1956]: 129–38).

8. Klaus Schmidt, "Göbekli Tepe—the Stone Age Sanctuaries: New Results of Ongoing Excavations with a Special Focus on Sculptures and High Reliefs," *Documenta Praehistorica*, no. 37 (2010): 239–56.

9. Jacques Cauvin, *Naissance des divinités, naissance de l'agriculture: la révolution des symboles au Néolithique* (Paris: CNRS, 1998).

10. E. B. Banning, "So Fair a House: Göbekli Tepe and the Identification of Temples in the Pre-pottery Neolithic of the Near East," *Current Anthropology* 52, no. 5 (2011): 619–60.

11. Heinrich von Kleist, *Kleist: Selected Writings*, trans. David Constantine (Indianapolis, Ind.: Hackett, 2004), 407.

12. Gilbert Simondon, "On Techno-Aesthetics," *Parrhesia* 14, no. 1 (2012): 1–8.

13. Dutton, *Art Instinct*, 13–28.

14. William Hirstein and Vilayanur Ramachandran, "The Science of Art: A Neurological Theory of Aesthetic Experience," *Journal of Consciousness Studies* 6, no. 6–7 (1998): 15–51.

15. Colin Martindale, *The Clockwork Muse: The Predictability of Artistic Change* (New York: Basic, 1990).

16. Dutton, *Art Instinct*.

17. The mechanism behind this progressive segregation of species in terms of the male's "costly" display of fitness through energetically less-efficient characteristics, such as extravagant plumage, is often referred to as "Fisherian Runaway."

18. Stephen J. Gould and Richard C. Lewontin, "The Spandrels of San Marco and the Panglossian Paradigm: A Critique of the Adaptationist Programme," *Proceedings of the Royal Society of London. Series B. Biological Sciences* 205, no. 1161 (September 21, 1979), 590.

19. Niles Eldredge and Stephen J. Gould, "Punctuated Equilibria: An Alternative to Phyletic Gradualism," in *Models in Paleobiology*, ed. Thomas J. M. Schopf (San Francisco: Freeman and Cooper, 1972), 82–115.

20. Henri Bergson, *Évolution Créatrice* (Saguenay, QC: J.-M. Tremblay, 2003).

21. Varela, Thompson, and Rosch, *Embodied Mind,* 191.

22. Niles Eldredge, "The Sloshing Bucket: How The Physical Realm Controls Evolution," in *Evolutionary Dynamics: Exploring the Interplay of Selection, Accident, Neutrality, and Function,* ed. James Crutchfield and Peter Schuster (Oxford: Oxford University Press, 2003), 3–32.

23. Humberto Maturana and Francisco J. Varela, *Autopoesis and Cognition: The Realization of the Living,* ed. Robert S. Cohen and Marx W. Watofsky, Boston Studies in the Phiosophy of Science 42 (Dordrecht and Boston: D. Reidel, 1980).

24. Varela, Thompson, and Rosch, *Embodied Mind,* 198.

25. Susan Oyama as quoted in Varela, Thompson, and Rosch, *Embodied Mind,* 199.

26. Gilles Deleuze and Félix Guattari, *A Thousand Plateaus: Capitalism and Schizophrenia* (Minneapolis: University of Minnesota Press, 1987), 10.

27. Deleuze and Guattari, *Thousand Plateaus,* 313.

28. Gould and Lewontin, "Spandrels of San Marco."

29. Hirstein and Ramachandran, "Science of Art," 16.

30. Marco Costa and Leonardo Corazza, "Aesthetic Phenomena as Supernormal Stimuli: The Case of Eye, Lip, and Lower-Face Size and Roundness in Artistic Portraits," *Perception* 35, no. 2 (2006): 236.

31. Hirstein and Ramachandran, "Science of Art," 18.

32. Hirstein and Ramachandran, "Science of Art," 16.

33. John Hyman, "Art and Neuroscience," in *Beyond Mimesis and Convention,* ed. Roman Frigg and Matthew Hunter, Boston Studies in the Philosophy of Science 262 (Dordrecht: Springer, 2010), 250.

34. David J. Buller, *Adapting Minds: Evolutionary Psychology and the Persistent Quest for Human Nature* (Cambridge, Mass.: MIT Press, 2005), 97.

35. Buller, *Adapting Minds,* 98.

36. Buller, *Adapting Minds,* 99.

37. Buller, *Adapting Minds,* 99.

38. Niklas Luhmann, *Art As a Social System,* trans. Eva M. Knodt (Stanford, Calif.: Stanford University Press, 2000), 49.

39. Luhmann, *Art as a Social System,* 253.

40. Richard O. Prum, "Coevolutionary Aesthetics in Human and Biotic Artworlds," *Biology & Philosophy* 28, no. 5 (2013): 811–32.

41. Richard O. Prum, "The Lande-Kirkpatrick Mechanism Is the Null Model of Evolution by Intersexual Selection: Implications for Meaning, Honesty, and Design in Intersexual Signals," *Evolution: International Journal of Organic Evolution* 64, no. 11 (November 2010): 3085–100.

42. Souriau, *L'avenir de l'esthetique,* 22–26.

43. Deleuze and Guattari, *Thousand Plateaus,* 316.

44. Gilles Deleuze and Félix Guattari, *What Is Philosophy?* European Perspectives (New York: Columbia University Press, 1994), 183.

45. Elizabeth Grosz, *Chaos, Territory, Art: Deleuze and the Framing of the Earth* (New York: Columbia University Press, 2008), 7.

46. Grosz, *Chaos, Territory, Art*, 11.
47. Deleuze and Guattari, *Thousand Plateaus*, 313.
48. Deleuze and Guattari, *Thousand Plateaus*, 314.
49. Deleuze and Guattari, *Thousand Plateaus*, 321.
50. Deleuze and Guattari, *Thousand Plateaus*, 316.
51. Jean-Marie Schaeffer, *Théorie des signaux coûteux, esthétique et art* (Rimouski: Presses de l'Université du Québec, 2010), 32–33 (my translation).
52. Deleuze and Guattari, *Thousand Plateaus*, 314.
53. Victor Turner, "Betwixt and Between: Liminal Period," in *The Forest of Symbols: Aspects of Ndembu Ritual* (Ithaca, N.Y.: Cornell University Press, 1970), 94.
54. Bernard Stiegler, *De la misère symbolique*, vol. 2, *La catastrophe du sensible* (Paris: Éditions Galilée, 2005), 154.
55. Gilles Deleuze, *Essays Critical and Clinical* (Minneapolis: University of Minnesota Press, 1997), 135.
56. Gilles Deleuze and Félix Guattari, *Qu'est-ce que la philosophie?* (Paris: Éditions de Minuit, 2005), 158.

3. Aesthesis and Prosthesis

1. Gilbert Simondon, *Du Mode d'existence Des Objets Techniques* (Paris: Aubier, 2001), 179–210.
2. Gilbert Simondon, *On the Mode of Existence of Technical Objects*, trans. Dan Melamphy, Nandita Biswas Melamphy, and Ninian Melamphy, Translation-thus-far (Independent, 2010), pt. 3 ("The Essence of Technicity"; rough draft), ch. 1 ("The Genesis of Technicity"), academia.edu/4184556/Gilbert_Simondon_On_the_Mode_of_Existence_of_Technical_Objects.
3. See "La vie, la mort," transcript of one of Derrida's unpublished seminars circulating online, code number DRR 173 (1/4).
4. Sylvain Auroux, *La révolution technologique de la grammatisation: introduction à l'histoire des sciences du langage* (Liège: Mardaga, 1994).
5. Henri Bergson, *Matière et Mémoire: Essai Sur La Relation Du Corps à l'esprit* (Saguenay, QC: J.-M. Tremblay, 2003), 17.
6. Simondon, *Mode of Existence*, 104.
7. Simondon, *Mode of Existence*, 99.
8. Friedrich Wilhelm Joseph von Schelling, *First Outline of a System of the Philosophy of Nature*, trans. Keith R. Peterson, Contemporary Continental Philosophy (Albany: State University of New York Press, 2004), 80.
9. Schelling, *First Outline*, 80.
10. Simondon, *Mode of Existence*, 125.
11. Schelling, *First Outline*, 80
12. Alfred North Whitehead, *Process and Reality* (New York: Free Press, 1982), 35.
13. Gilles Châtelet, *Figuring Space: Philosophy, Mathematics, and Physics*, Science and Philosophy 8 (Dordrecht and Boston: Kluwer, 2000), 50.

14. Francis Bailly and Giuseppe Longo, *Mathematics and the Natural Sciences: The Physical Singularity of Life*, Advances in Computer Science and Engineering: Texts 7 (London and Hackensack, N.J.: Imperial College Press, 2011), 76.

15. Alfred North Whitehead, *Science and the Modern World*, reissue ed. (New York: Free Press, 1997), 213.

16. William James, *The Principles of Psychology*, vol. 1 (New York: Dover, 1890), 404.

17. For several attempts at this, see *Naturalizing Phenomenology: Issues in Contemporary Phenomenology and Cognitive Science*, ed. Jean Petitot et al. (Stanford, Calif.: Stanford University Press, 2000).

18. William James, *The Meaning of Truth : A Sequel to "Pragmatism"* (London: Longmans, Green, 1909), 116.

19. Samuel Butler, *Life and Habit* (London: Trübner, 1878), 18.

20. Gilles Deleuze, *Difference and Repetition* (New York: Columbia University Press, 1994), 71.

21. Michael Kirchhoff et al., "The Markov Blankets of Life: Autonomy, Active Inference, and the Free Energy Principle," *Journal of The Royal Society Interface* 15, no. 138 (2018): royalsocietypublishing.org/doi/full/10.1098/rsif.2017.0792.

22. Gilbert Simondon, *L'individuation à La Lumière Des Notions de Forme et d'information* (Paris: Jérôme Millon, 2005), 228 (my translation).

23. Humberto R. Maturana and Francisco J. Varela, *Autopoiesis and Cognition: The Realization of the Living* (Dordrecht and Boston: D. Reidel, 1980).

24. Simondon, *L'individuation*, 228 (my translation).

25. Georg Wilhelm Friedrich Hegel, *Hegel's Philosophy of Mind*, trans. William Wallace (Oxford : New York, 2010), 131.

26. Edmund Husserl, *On the Phenomenology of the Consciousness of Internal Time (1893–1917)*, vol. 4 of *Edmund Husserl: Collected Works*, trans. John B. Brough (Dordrecht and Boston: Kluwer Academic, 1991), 49–50.

27. Hegel, *Philosophy of Mind*, 131.

28. Deleuze, *Difference and Repetition*, 79.

29. Deleuze, *Difference and Repetition*, 80.

30. Brian Massumi, "Envisioning the Virtual," in *The Oxford Handbook of Virtuality*, ed. Mark Grimshaw (New York: Oxford University Press, 2014), 61.

31. Gilles Deleuze and Claire Parnet, *Dialogues* (Paris: Flammarion, 2008), 184.

32. Simondon, *L'individuation*, 228 (my translation).

33. Simondon, *L'individuation*, 228 (my translation).

34. Giuseppe Longo and Maël Montévil, *Perspectives on Organisms: Biological Time, Symmetries, and Singularities*, Lecture Notes in Morphogenesis (Berlin and Heidelberg: Springer, 2014).

35. Gilles Deleuze, *Différence et Répétition* (Paris: Presses Universitaires de France, 2000), 102.

36. Oliver Sacks, "The Abyss." *The New Yorker*, September 24, 2007, newyorker.com/magazine/2007/09/24/the-abyss.

37. Sacks, "Abyss," 4.

38. Heinz von Foerster, "Ethics and Second-Order Cybernetics," in *Understanding Understanding: Essays on Cybernetics and Cognition* (New York: Springer, 2003), 301.

39. Bernard Stiegler, *Technics and Time*, vol. 2, trans. Stephen Barker (Stanford, Calif.: Stanford University Press, 2008), 37.

40. Plato, *Plato: Laches, Protagoras, Meno, Euthydemus*, trans. W. R. M. Lamb (Cambridge, Mass.: Harvard University Press, 1977), 303.

41. See Eric R. Kandel, *In Search of Memory: The Emergence of a New Science of Mind* (New York: Norton, 2007).

42. Bernard Stiegler, *Philosopher Par Accident : Entretiens Avec Elie During* (Paris: Galilée, 2004), 216.

43. Bernard Stiegler, "Memory," in *Critical Terms for Media Studies*, ed. W. J. T. Mitchell and Mark B. N. Hansen (Chicago: University of Chicago Press, 2010), 67.

44. Stiegler, "Memory," 80.

45. Chris R. Reid et al., "Slime Mold Uses an Externalized Spatial 'Memory' to Navigate in Complex Environments," *Proceedings of the National Academy of Sciences* 109, no. 43 (October 23, 2012): 17490–94, at 17490 (emphasis added).

46. Stiegler, *Technics and Time*, 2:167.

47. Per Bak, *How Nature Works: The Science of Self-Organized Criticality* (New York: Copernicus, 1996).

48. Deleuze, *Différence et Répétition*, 108–9.

49. Luciana Parisi, *Abstract Sex: Philosophy, Bio-Technology, and the Mutations of Desire*, Transversals (London and New York: Continuum, 2004), 60.

50. See W. R. Ashby, "Principles of the self-organizing system," in *Principles of Self-Organization Transactions of the University of Illinois Symposium on Self-Organization, Robert Allerton Park, 8 and 9 June, 1961*, ed. Heinz Von Foerster and George W Zopf (Oxford: Pergamon, 1962). 255–78.

51. Stiegler, *Philosopher Par Accident*, 48.

52. Karl Friston, James Kilner, and Lee Harrison, "A Free Energy Principle for the Brain," *Journal of Physiology-Paris* 100, no. 1–3 (2006): 70–87.

53. Friston, Kilner, and Harrison, "Free Energy Principle", 72.

54. Jacques Derrida, *Of Grammatology*, trans. Gayatri Chakravorty Spivak (Baltimore, Md.: Johns Hopkins University Press, 1998), 143.

55. Stiegler, *Technics and Time*, 2:135.

56. Whitehead, *Process and Reality*, 73.

57. Robert B. Brandom, *Making It Explicit: Reasoning, Representing, and Discursive Commitment* (Cambridge, Mass.: Harvard University Press, 1998).

58. David Roden, *Posthuman Life: Philosophy at the Edge of the Human* (London: Routledge, 2015).

59. Franscisco J. Varela, Evan T. Thompson, and Eleanor Rosch, *The Embodied Mind: Cognitive Science and Human Experience* (Cambridge, Mass: MIT Press, 1992), 181–83.

4. Superposition and Time

1. Quentin Meillassoux, *Time without Becoming* (Sesto San Giovanni, Italy: Mimesis Edizioni, 2014), 11.

2. Meillassoux, *Time without Becoming,* 13.

3. Rick Dolphijn and Iris van der Tuin, *New Materialism: Interviews and Cartographies* (Ann Arbor: University of Michigan Library, 2012), 81.

4. Meillassoux, *Time without Becoming,* 23.

5. Quentin Meillassoux and Alain Badiou, *After Finitude: An Essay on the Necessity of Contingency,* trans. Ray Brassier (London and New York: Bloomsbury Academic, 2010), 87 (emphasis added).

6. Michel Bitbol has since published a book explicitly outlining his defense of correlationism (*Maintenant la finitude: Peut-on penser l'absolu?* [Paris: Flammarion, 2019]).

7. Michel Bitbol, "Reflective Metaphysics: Understanding Quantum Mechanics from a Kantian Standpoint," *Philosophica* 83 (2008): 53–83.

8. Michel Bitbol, *De l'intérieur du monde: Pour une Philosophie et une science des relations,* Bibliothèque Des Savoirs (Paris: Flammarion, 2010).

9. Aristotle, *Posterior Analytics* 1.3, trans. G. R. G. Mure (Adelaide, Australia: University of Adelaide, 2015), ebooks.adelaide.edu.au/a/aristotle/a8poa/.

10. Aristotle, *Posterior Analytics* 1.6.

11. G. W. Leibniz, *G. W. Leibniz's Monadology,* trans. Nicholas Rescher (Pittsburgh, Pa.: University of Pittsburgh Press, 1991), §§ 37, 21–22.

12. Steven Weinberg, *Dreams of a Final Theory: The Scientist's Search for the Ultimate Laws of Nature,* repr. ed. (New York: Vintage, 1994), loc. 3384.

13. John Archibald Wheeler, *Information, Physics, Quantum: The Search for Links* (Austin: Physics Department of the University of Texas, 1990), 314.

14. For more on this topic, see Alexander Wilson, "Big Data and the Thermodynamics of Discretisation," *London Journal of Critical Thought* 1, no. 1 (2016): 61–72.

15. Quentin Meillassoux, *After Finitude: an Essay on the Necessity of Contingency,* with preface by Alain Badiou (London : Continuum, 2008).

16. Immanuel Kant, *Critique of Judgment,* trans. Werner S. Pluhar (Indianapolis, Ind.: Hackett, 1987), 252.

17. Richard Phillips Feynman, *The Feynman Lectures on Physics,* vol. 1, *Mainly Mechanics, Radiation, and Heat* (Reading, Mass.: Addison-Wesley, 2008), 47–48.

18. Lewis Carroll, "What the Tortoise Said to Achilles," *Mind* 4, no. 14 (1895): 278–80.

19. Pierre Simon Laplace, *A Philosophical Essay on Probabilities* (New York and London: J. Wiley and Chapman & Hall, 1902), 4.

20. Roberto Mangabeira Unger and Lee Smolin, *The Singular Universe and the Reality of Time: A Proposal in Natural Philosophy* (New York: Cambridge University Press, 2014), 347.

21. Charles Pierce, "The Architecture of Theories," *The Monist* 1, no. 2 (1891): 165: "Law is *par excellence* the thing that wants a reason."

22. Unger and Smolin, *Singular Universe,* 367.

23. Note that Smolin's appropriation of Leibniz mischaracterizes the principle of sufficient reason somewhat. For, as noted earlier in the chapter, Leibniz argues that the sufficient reason lies "outside" the series of infinite contingencies. See Unger and Smolin, *Singular Universe,* 367.

24. Unger and Smolin, *Singular Universe,* 440.

25. The cutoff point seems to be somewhere around Avogadro's number (6.02 x 1010), which approximates the distinction between molar and molecular.

26. Carlo Rovelli, *Et si le temps n'existait pas?: un peu de science subversive* (Paris: Dunod, 2014), 118.

27. Gilles Deleuze, *Différence et Répétition* (Paris: Presses Universitaires de France, 2000), 270.

28. Karen M. Barad, "Diffracting Diffraction: Cutting Together-Apart," *Parallax* 20, no. 3 (July 3, 2014): 168–87.

29. Karen M. Barad, *Meeting the Universe Halfway: Quantum Physics and the Entanglement of Matter and Meaning* (Durham, N.C.: Duke University Press, 2007).

30. Gilles Deleuze, *Difference and Repetition,* trans. Paul Patton (New York: Columbia University Press, 1994), 223.

31. Deleuze, *Difference and Repetition,* 224.

32. Gilbert Simondon, *On the Mode of Existence of Technical Objects,* trans. Dan Melamphy, Nandita Biswas Melamphy, and Ninian Melamphy, Translation-thus-far (Independent, 2010), pt. 3 ("The Essence of Technicity"; rough draft), ch. 1 ("The Genesis of Technicity"), academia.edu/4184556/Gilbert_Simondon_On_the_Mode_of_Existence_of_Technical_Objects.

33. Friedrich Wilhelm Joseph von Schelling, *First Outline of a System of the Philosophy of Nature,* trans. Keith R. Peterson, SUNY Series in Contemporary Continental Philosophy (Albany: State University of New York Press, 2004), 105.

34. Arturo Rosenblueth, Norbert Wiener, and Julian Bigelow, "Behavior, Purpose, and Teleology," *Philosophy of Science* 10, no. 1 (1943): 18–24.

35. Gilbert Simondon, *L'individuation à la lumière des notions de forme et d'information* (Paris: Jérôme Millon, 2005), 160.

36. Benoit B Mandelbrot, *The Fractal Geometry of Nature* (New York: W. H. Freeman, 1983).

37. Francis Bailly and Giuseppe Longo, "Causes and Symmetries in Natural Sciences," 2004, ftp://129.199.99.98/pub/users/longo/CIM/cause-symMoreGeo.pdf.

38. Ilya Prigogine and Isabelle Stengers, *Order Out of Chaos* (New York: Bantam, 1984).

39. Alfred North Whitehead, *Science and the Modern World,* repr. ed. (New York: Free Press, 1997), 103.

40. Whitehead, *Science and the Modern World,* 104.

41. Whitehead, *Science and the Modern World,* 87.

42. Brian Massumi, "Virtual Ecology and the Question of Value," in *General Ecology: The New Ecological Paradigm,* ed. Erich Hörl (London and New York: Bloomsbury Academic, 2017).

43. Alfred North Whitehead, *Process and Reality* (New York: Free Press, 1982), 43.

44. Félix Guattari, *The Three Ecologies,* trans. Ian Pindar and Paul Sutton (London: Bloomsbury Academic, 2008), 45.

45. Gilbert Simondon, *Imagination et invention* (Chatou: Éditions de la Transparence, 2008), 91 (my translation).

46. This account resonates with R. Scott Bakker's "blind brain theory" and is inseparable from the problem of the "ineffability" of *qualia*; see R. S. Bakker, "The Last Magic Show: A Blind Brain Theory of the Appearance of Consciousness," April 2012, academia.edu/1502945/The_Last_Magic_Show_A_Blind_Brain_Theory_of_the_Appearance_of_Consciousness.

47. Giulio Tononi, "Consciousness as Integrated Information: A Provisional Manifesto," *The Biological Bulletin* 215, no. 3 (December 2008): 216–42.

48. Wojciech H. Zurek, "Decoherence and the Transition from Quantum to Classical—REVISITED," *Los Alamos Science,* no. 27 (2002): 4, arxiv.org/ftp/quant-ph/papers/0306/0306072.pdf (updated from *Physics Today* 44 [1991]: 36–44).

49. Masafumi Oizumi, Larissa Albantakis, and Giulio Tononi, "From the Phenomenology to the Mechanisms of Consciousness: Integrated Information Theory 3.0," *PLoS Comput Biol* 10, no. 5 (May 8, 2014): 3.

50. Oizumi, Albantakis, and Tononi, "From the Phenomenology to the Mechanisms," 3.

51. Indeed, even David Chalmers has shown tentative support for Integrated Information Theory's potential.

52. Christof Koch, *Consciousness: Confessions of a Romantic Reductionist* (Cambridge, Mass.: MIT Press, 2012).

53. John R. Searle, "Can Information Theory Explain Consciousness?," *The New York Review of Books,* January 10, 2013, nybooks.com/articles/archives/2013/jan/10/can-information-theory-explain-consciousness/.

54. Scott Aaronson, "Why I Am Not an Integrated Information Theorist (or, 'The Unconscious Expander')," Shtetl-Optimized (blog of Scott Aaronson), May 21, 2014, scottaaronson.com/blog/?p=1799.

55. Max Tegmark, "How Unitary Cosmology Generalizes Thermodynamics and Solves the Inflationary Entropy Problem," *Physical Review D* 85, no. 12 (June 11, 2012), dspace.mit.edu/openaccess-disseminate/1721.1/72144.

56. Barad, "Diffracting Diffraction."

57. Deleuze, *Difference and Repetition,* 212.

58. It is from this perspective that Deleuze can claim that, of the two elements of a repetitive series, neither can be said to be strictly original or derived. For, the former and the actual instantiate two parallel series as a "function of the virtual object." All actualizations, even the present that has now passed, instantiate their own relative subjectivities. For Deleuze, the former present is subordinated to a principle of *identity,* while the actual present is subordinated to one of *resemblance,* marking the familiar components of repetition. But for him, it is the virtual object that ensures the constant articulation of these two modes and that subsists, therefore, as an amodal translation from one set of symmetries to another. Thus, a single par-

tial object, in its virtual plunge, constitutes a particular transduction between two regimes, via a *quasi-causal* transduction. It is in this sense that the virtual object "n'est trouvé que comme perdu—il n'existe que comme retrouvé." For, insofar as the virtual object is a fragment of pure past, its appearance is *contingent,* and thus always invokes an *anamnesis* of such a quasi origin.

Conclusion

1. Alfred North Whitehead, *Science and the Modern World,* repr. ed. (New York: Free Press, 1997), 201.

INDEX

Aaronson, Scott, 199
Abramovich, Roman, 69
Actaeon, 132
actor-network theory, 2
actual, 50, 54; genesis of, 203; occasion, 152, 154, 188; products of, 172; virtual and, 204
adaptation, 85, 221n7; importance of, 94; prototechnical, 144
adaptationists, 69, 71, 82; aesthetics and, 91
Adorno, Theodor, 106
aesthesia, 8, 48, 143; as beyond the human, 198; genesis of, 186; independence of, 212; nontotalizable, 116
aesthesic integration, 110, 198
aesthesis, 1, 41, 49–50, 152, 170; asymptote of, 209–10; coarse-graining of nature and, 15; cognition and, 47, 66; discretization and, 114; material subjectivities and, 207; memory and cognition and, 109; necessary truths and, 42; nonhuman or alien, 11; perceptronium and, 192; prejudices against, 79; process of, 95, 115; symmetry breaking, 178; territorialization and, 101; unconditioned nature of, 49
aestheta, 10, 26; *noeta* and, 23, 27, 43, 46, 79; pure, 65

aesthetic reticulation, 14
aesthetics, 10, 17–18; academic, 70; adaptationists and, 91; Aristotle on, 68; art and, 71; Baumgarten and, 19, 24, 25, 40; cognition and, 80; depragmatized moment, 102; dismissal of, 29; environmental pressures and, 97; epistemics and, 47; evolutionary, 82; foundation of, 27; judgment and, 110, 167; Kant and, 68; Leibniz and, 97; neuro-, 79; nonhuman, 7, 12, 214; philosophy and, 23; preferences, 67; reinvention of, 42; schemata, 75; sexual selection and, 96; Simondon on, 13, 113; taste and, 24; transcendental, 35
aesthetic selection, 12, 97, 99
African divination, 62
Aionic temporality, 190
America, 65
amnesia, 123; organisms and, 127; technologically induced, 139
amnesis, 139, 151
analyticity, 38
analytic reason, 29
analytic–synthetic distinction, 65, 138
anamnesis, 124, 139, 151
Anderson, P. W., 116
animal art, 105
animal sentience, 1, 152

animal territorializations, 98
anisotropic, 182
Anthology of Chance Operations, An, 52
Anthropocene, 1, 154, 211, 214; effects
    of, 7
anti-anthropocentric stance, 3
antirealism, 157
aperiodic behavior, 62
Après la finitude, 155
arche-fossil, 156
arche-trace, 142
Aristotle, 39, 67, 75, 120, 164; on aesthetic
    practices, 68
Arnaud, Antoine, 177
Arp, Jean, 51
arrow of logical priority, 170
art, 50, 106–7; aesthetics and, 71; animal,
    105; animal territorializations and,
    98; artistic exteriorization, 109;
    coefficient, 52; cult of genius, 53;
    deep structure of, 89, 90; dismissal
    as frivolous, 65, 67; epistemaesthetic
    realization of, 46; evolution and, 81;
    free time and resources for, 68, 73;
    material causes of, 72; materialist
    and idealist perspectives on, 76–78;
    as ostentation, 66, 71; outcome of
    artistic event, 51; outsider, 69–70;
    purposelessness of, 102, 106; science
    and, 45; sexual selection and, 69, 81;
    symmetry-breakings and, 54; will to,
    77, 78
artificial intelligence, 2
artistic exteriorization, 98, 109
association, 145
a-territorial liminality, 106
Augustine, 109, 120, 121, 179
aura, 64
Auriti, Marino, 69
Auroux, Sylvain, 112
autopoiesis, 86
autopoietic: being, 167; loop, 186; unit,
    160

autoreflection, 159
avant-garde, 51
awareness, 122

Badiou, Alain, 17
Bailly, Francis, 38, 145
Bak, Per, 144
Bamana divination, 54, 61
Banning, E. B., 73
Barad, Karen, 178, 202
base-2 counting scheme, 56
Bateson, Gregory, 196
Baumgarten, Alexander Gottlieb, 4, 10;
    aesthetics and, 19, 24, 25, 40; Kant
    and, 23
beauty, 62
Being and Time, 39
Benjamin, Walter, 64
Bergson, Henri, 84, 113, 129, 173; critique
    of the possible, 118
best-of-possible-worlds doctrine, 201
binary number system, 56
biology, 184
bio-resonance, 58
Bitbol, Michel, 14, 35, 159–62, 166, 168
Bolyai, János, 30
Book of Changes, 56
boundary, 33, 47
Bourdieu, Pierre, 68
Bouvet, Joachim, 56
Brentano, Franz, 121, 127
Breton, André, 51
Brunelleschi, Filippo, 75
Buiatti, Marcello, 58
Buller, David, 90
Burroughs, William S., 51
Butler, Samuel, 123
Byrkit, James Ward, 194

Cage, John, 51
Cantor, Georg, 9
Carnap, Rudolf, 9, 34
Carroll, Lewis, 36, 171

"Cartesian theatre," 3
Cassirer, Ernst, 160
catharsis, 68
Catren, Gabriel, 7
Cattelan, Maurizio, 65
causality, 15, 32, 96, 124; circular, 186; explanations for, 163; Hume and, 125; mechanism of, 181; physics and, 169; undifferentiated whole of, 54. *See also* quasi-causal(ity)
causation, 17, 173
cause-effect information, 196
Cauvin, Jacques, 73
Cézanne, Paul, 45
Chaitin, Gregory, 62
Chalmers, David, 195
chance, 51, 54, 56–57, 106, 173–74
chaoïd, 106, 107
chaos, 15, 17; deterministic, 55, 57–58, 61; hyper-, 158, 161, 163, 167
chaosmos, 16, 155, 205, 206
Châtelet, Gilles, 117
Church, Alonzo, 56; Typed Lambda Calculus, 60
Churchland, Paul, 4
civilization, 72, 73, 113
classical dynamics, 20
clinamen, 115, 183, 184, 185
coarse-grained objects, 175
coarse-graining of nature, 15, 172, 177
cognition, 34, 109, 190; aesthesis and, 47, 66, 80; conditions of, 31; constraints on, 8, 48; distinct, 43; distinction and, 23; elementary cognitive subsystems, 133; foundation of, 60; of intrinsically confused knowledge, 23; kinds of, 24; material transcendental conditions of, 118; memory and, 118; *qualia* and, 191; Simondon on, 126; symmetry breaking and, 54; synthesis and, 50. *See also* higher cognitive faculties; lower cognitive faculties
Coherence (Byrkit), 194

collective consciousness, 200
commedia dell'arte, 52
compossibility, 48, 172, 200; provisional compossible world, 210; sieve of, 209; web of, 207
computability, 63
computational equivalence, 58, 62
computers, 55
computer science, 16
computronium, 197–98
concrescence, 114
conditions: a priori, 34, 105, 137, 161, 214; of cognition, 31; conditioning factors, 7; implicit, 137; of intelligibility, 10; of perceptronium, 8; of real experience, 5; transcendental, 190, 191
Confessions, 120
confusion, 21, 23
Connes, Alain, 176
consciousness, 17, 120, 196; collective, 200; hard problem of, 195, 197; as integrated information, 192
constructive interference, 94
constructive type theory, 60
continental philosophy, 38
contingency, 137, 164, 168, 169, 184; caring for, 52; necessity and, 48; unpredictability and, 53
continuous, 26, 33, 59, 61, 110–11; catastrophe, 104; discrete and, 50; sequence of maturation, 103
Copenhagen interpretation, 179
Copernican doctrine, 154
correlationism, 6–7, 155, 159; Meillassoux on, 14; naïve, 214
cosmic inflation, 201
critical posthumanism, 2
Critique of Pure Reason, 157
cubism, 46
Cudworth, Ralph, 28–29
cult of genius, 53
Curry-Howard isomorphism, 60

cutting together-apart, 178
cyborg feminism, 2

Dada, 51
daemon, 52
Dalí, Salvador, 46
Darwin, Charles, 12, 81, 83, 84
decoherence, 17, 192–95, 200–202
deconstruction, 40
degrees of freedom, 151
delayed-choice experiment, 179
Deleuze, Gilles, 5, 9, 13, 37, 141; Aionic
    temporality and, 190; on art, 50,
    106–7; on deterritorialization, 106; on
    diversity, 181; early master works of,
    205; on emergence of language, 31;
    on evolution, 86–87; on individual
    and totality, 204; Kant and, 48–49;
    Leibniz and, 23; on ontologies of
    identity, 118; paradoxes and, 42;
    passive synthesis and, 32, 124; on
    philosophy, 7; quasi-causality and,
    16, 182; on second synthesis of time,
    128, 129; on static genesis, 158; on
    territorialization, 99, 100, 101, 111;
    on third critique, 168; transcendental
    empiricism, 34, 180; on transcoding,
    100; on virtual, 131
Deller, Jeremy, 69
demarcation, 9
Democritus, 29
Dennett, Daniel, 3, 59
depragmatized aesthetic moment, 102
Derrida, Jacques, 9, 137, 140; on arche-
    trace, 142; Heidegger and, 40; Jacob
    and, 112; metaphysics of presence
    and, 39; mnemotechnics and, 13;
    Origin of Geometry, 136; on supple-
    ments, 150, 180; trace and, 38; on
    writing, 139, 141
Descartes, René, 16, 19, 27, 177
determinism, 173
deterministic chaos, 55, 57–58, 61
deterministic constraints, 53

deterritorialization, 104, 106
development, 84
dice, 56–57
différance, 40
difference, 141
Difference and Repetition, 205
Dirac, Paul, 75
disconnection thesis, 153
discrete, 50, 61, 112, 121
discretization, 111–15
discrimination learning, 89
displacement, 199
distinction, 22, 23, 41, 212
diversity, 181
divinatory practices, 56
divine will, 52
division of labor, 113
dogmas of empiricism, 38
doxa, 28, 69, 70
drift, 83, 84, 92, 94, 95
dualism, 27
Duchamp, Marcel, 51, 52
Durkheim, Émile, 68
Dutton, Denis, 70, 80

Easter Island, 76
Eglash, Ron, 55
Einstein, Albert, 34, 58, 75, 160, 173
Eldredge, Niles, 85
elementary cognitive subsystems, 133
Elements of Drawing, The, 45
eliminativism, 1–2, 211; obstacles faced
    by, 3; panpsychism and, 4
emergence, 173
emergent phenomena, 147
enabling constraints, 126
enaction, 86, 87
energy, 71
entanglement, 151, 178, 189, 194, 208;
    among aspects of cognition, 34; of
    knowledge and sensation, 8; of little
    perceptions, 201; locality and, 17, 202;
    nested, 206; of observer and world,
    165; of organism and world, 11, 180;

quasi-causality, 207; realization of, 44

entropy, 15

Entscheidungsproblem (decision problem), 56

environments: aesthetics and, 97; distinction between organisms and, 85–86; environmental selection, 71, 93

epiphylogenesis, 143

epistemaesthetics, 5, 43, 79, 95–96, 154; categories, 48; convergences, 78; desires, 75; material emergence and, 207; production of intelligibility and, 103; realization, 46

epistemics: aesthetics and, 47; biases, 45; holism, 39; relativism, 155

epoché, 79, 211

eternal objects, 188, 206

Euclid, 30

evental adsorption, 152

Everett, Hugh, 193

evolution, 27, 76; aesthetics, 82; arms race, 91–92, 93; Deleuze and Guattari on, 86–87; selection, 148; survival-affording responses and, 88; *technē* and, 79

evolutionary biology, 93

exclusion, 1, 202

experience, 123, 151–52; aestheta and, 10, 26; constraints on forms of, 1, 8; experiential integration, 200; "folk" notions of, 3; incomplete naturalization of, 214; necessary conditions of, 6; possible, 5; superject and, 187; traditional notions of, 4

extended critical systems, 13

exteriorization, 14, 98–99, 109; indistinction and, 148; mnemotechnical, 106, 142, 147; retentional, 145; variants of, 110

family rhizome, 87

feedback loops, 86, 91, 92, 99

Fertile Crescent, 72

Feynman, Richard, 30, 170

Fichte, Johann Gottlieb, 157

Fisher, Ronald, 12, 96

fitness, 148

Floridi, Luciano, 61

fluctuation, 78

Fluxus movement, 51, 52

Foerster, Heinz von, 136

For the Love of God, 65

fossil record, 84

foundationalism, 4, 164; philosophy and, 37–38; reductionism, 38; Sellars and, 39

frame problem, 2

Friedrich, Caspar David, 63

Friston, Karl, 149, 150

functional programming languages, 16

fundamental physics, 164, 171–72

Fu Xi, 56

Galilei, Galileo, 29

Gauss, Carl Friedrich, 30

general organology, 137, 141, 205

genetic drift, 83

geodesics, 32, 60, 186

geometry, 30–32, 54

German idealism, 157

Gestell (framing), 99–100, 101

Gettier, Edmund, 43, 44, 46

Gioni, Massimiliano, 69

givenness, 4, 39, 66

Göbekli Tepe, 11, 72, 73, 74, 76, 81

God, 21–22, 120, 158

Gödel, Kurt, 9, 56; incompleteness theorems of, 117

Gombrich, Ernst, 45, 78

Gould, Stephen Jay, 12, 70, 84, 87

grammatization, 112, 115

great outdoors, 155, 157, 180; Bitbol on, 161; as hyperchaos, 163; Meillassoux on, 162

Greek culture, 52, 67; arts and, 66; Kant and, 24; knowledge and, 43; philosophy and, 23

Grosz, Elizabeth, 13, 98
Guattari, Félix, 13; on art, 50, 106–7; on
deterritorialization, 106; on evolution,
86–87; on territorialization, 99, 100,
101, 111; on transcoding, 100

habit, 122–24, 130, 140; contraction
of events as, 146; entrenched, 32;
essential nature of, 126; established,
128; formation, 125, 133, 136; Hegel
and, 127, 129; Hume and, 31, 167;
mathematics and, 33; process of, 181;
synthesis of, 141
habitus, 69, 137
haecceity, 190
Hamiltonian, 201, 203, 204
handicap principle, 83
heat, 3
heat death, 180
Hegel, Georg Wilhelm Friedrich, 29, 127,
129, 157
Heidegger, Martin, 9, 39, 99, 122; Derrida
and, 40; originary technicity and, 137;
on technical being, 127
Heisenberg, Werner, 176
Helmholtz, Hermann von, 169
Heraclitus, 158, 208
hidden information, 15
higher cognitive faculties, 42, 59–60
Hilbert, David, 56
Hilbert Space, 161, 203, 204
Hirst, Damien, 65
Hirstein, William, 80
Hobbes, Thomas, 26
homeostasis, 122, 147
Homotopy-Type Theory, 60
human exceptionalism, 43
human sapience, 1, 152
human–world relation, 167
Hume, David, 31, 32, 125, 167, 177
hunter-gatherer societies, 72
Husserl, Edmund, 9, 13–14, 39, 121, 136;
on secondary memory, 128
hybridization, 215

hyperchaos, 158, 161, 167; great outdoors
as, 163
hypomnemata, 121, 136, 137, 143, 142

I-Ching divination practice, 51, 56
idealism, 77, 157
idealizations, 204
IIT. *See* integrated information theory
illusionism, 44, 45
immanence, 191
implicitation, 101
impressionism, 45–46
inclusion, 1, 2
incompleteness theorems, 38, 117
incomputable, 9, 16
inconsistent multiplicity, 116
indeterminacy, 205
Indian Chola sculptures, 89
indiscriminate pluralism, 2
indistinction, 22, 41, 97, 105, 212; exteriorization and, 148; Kant and, 25
individuation(s), 113 technosocial, 110;
technospiritual, 110
ineffability, 218n16; of *qualia*, 21, 24, 63
infinitesimal, 20–21, 58, 104; fractional
dimensions, 187; isomorphism, 26;
subdivisions, 117; time scales, 128
information, 15, 78, 166; cause-effect,
196; consciousness as integrated,
192; integration of, 154; matter-
information-observation loop, 165;
mutual, 195; Shannon's theory of, 169
informativity, 196
integrated information theory (IIT), 195–
96, 198–200, 202
integration, 109, 196; aesthesic, 110, 198;
experiential, 200; of information, 154;
maximum value of, 197; secondary,
128, 132, 133; tertiary, 136
intelligence, 1, 211–12
intelligibility, 10, 43, 104, 117, 178;
conditions of, 49; epistemeasthetic
production of, 103; fabric of, 115
interaction, 130

intercession, 114, 146; supersession and, 13, 14, 105, 144, 202, 208
interest of reason, 49
inter-structurality, 103
intraperceptive images, 190
irrational numbers, 25
irreducibility, 173

Jacob, François, 112
James, William, 47, 121, 123
je ne sais quoi, 25–26, 64; of experience, 10; of *qualia*, 21
judgment, 24, 110, 167
justified true belief, 43
just-so, 12, 70, 76, 88, 94

Kant, Immanuel, 5, 16, 166, 168, 204; aesthetics and, 68; Baumgarten and, 23; Deleuze and, 48–49; indistinction and, 25; Leibniz and, 24, 26; Maimon and, 31, 33, 163; Meillassoux and, 156, 157; neo-Kantians, 34; on purposeful purposelessness in art, 102; sublime and, 62
Kipling, Rudyard, 12
Kleene, Stephen, 60
Kleist, Heinrich von, 76
knowledge, 1, 5, 42; by acquaintance, 38; Aristotle on, 164; background of, 122; boundaries of experimental, 160; cognition of intrinsically confused, 23; construction of, 137; dimensions of, 22; genesis, 163; Hume and, 31; obscurity and, 20; as opposition to senses, 8; perception and, 44; philosophy and, 11; sensation and, 28, 30. *See also* objects of knowledge
Koch, Christof, 198, 199
kosmos, 206
Kuhn, Thomas, 104
Kunstwollen, 77, 78

language, 31
Laplace, Pierre-Simon, 15, 173, 176

Leibniz, Gottfried Wilhelm, 10, 17, 19, 33–34, 164; aesthetics and, 97; base-2 counting scheme and, 56; best-of-possible-worlds doctrine, 201; compossibility and, 48; computers and, 55; Deleuze and, 23; God and, 21–22; Hobbes and, 27; inconsistency in reasoning of, 21; Kant and, 24, 26; objects of knowledge and, 20; observations of, 213; principle of sufficient reason, 174; *qualia* and, 59; Smolin and, 227n23
leisure activities, 68
Leroi-Gourhan, André, 137, 142
Lewontin, Richard, 84, 87
Life and Habit, 123
liminal spaces, 103
Llull, Ramon, 55
Lobachevski, Nikolai, 30
Locke, John, 29, 47
logical empiricism, 34
logical positivism, 34, 38
logical time, 176–77, 178, 182, 186
Logic of Sense, 205
Longo, Guiseppe, 32, 38, 58, 145
loops, 185–86
Lorenz, Konrad, 101
lower cognitive faculties, 42, 59–60, 79
Luhmann, Niklas, 94

Maciunas, George, 51
macromutation, 84
magical realm, 114
Maimon, Solomon, 5, 34; Kant and, 31, 33, 163; relativism and, 35
Markov blankets, 125, 131
Martindale, Colin, 80
Martin-Löf, Per, 60
Marxism, 77
Massumi, Brian, 130, 189
material emergence, 207
materialism, 208
material semiotics, 2
mathematics, 16, 33, 46, 174; absolute

mathematical language, 55; constructivistic interpretations of, 44; internal pressures of, 32; irrational numbers and, 25; mathematical infinitesimal, 20; mathematical objects, 60; Platonist interpretation of, 175

matter, 4, 165–67, 183, 209; constitution of, 118; creativity and, 79; divisible, 27; entangled nature of, 5; exotic states of, 192, 197; as fluctuation, 78; mind and, 2, 171, 195, 210; new materialisms and, 11–12; organization of inorganic, 138; potential behaviors of, 199; principle of variation of, 211; provisionally organized, 212; reactivity of, 86; Schelling and, 116; states of, 127, 204; totality of all, 200; ungrounding of, 17

Matter and Memory, 113

Maturana, Humberto, 86

maximally irreducible cause-effect repertoire (MICE), 197

Meillassoux, Quentin, 16, 155–60, 163, 167; correlationism and, 14; on great outdoors, 162; Hamiltonian and, 203

memory, 109, 141, 151; amnesia, 123, 127, 139; cognition and, 118; episodic, 133, 134; Husserl on secondary, 128; organisms and, 119; procedural, 123, 130, 135; redistribution of, 140; semantic, 133, 134

Meno, 138, 140

mereological bio-resonance, 144

metaphysics, 39, 40, 159

metastability, 181, 187

Metzinger, Thomas, 24, 25

MICE. *See* maximally irreducible cause-effect repertoire

Michelangelo, 53

microscopic phenomena, 117

mnemotechnical assemblage, 130

mnemotechnical exteriorization, 106, 142, 147

mnemotechnics, 13

mnesis, 139, 151

monadology, 201

monads, 191–92

monism, 173

monuments, 74

Morris, William, 69

Morse, Marsten, 62

mutual presupposition, 87

Nagel, Thomas, 3

Naturphilosophie, 167

necessary conditions, 6

necessary truths, 42

necessity, 168, 203

negative feedback, 187

negentropic, 132, 149

neuro-aesthetics, 79

new materialisms, 11

Newton, Isaac, 33, 34, 175

niche construction, 91, 92

noeta: aestheta and, 23, 27, 43, 46, 79; pure, 65

noetic act, 50

non-Euclidean geometry, 31

nonhuman aesthetics, 7, 12, 214

nonhuman entities, 1

nonlinearity, 173

non-Platonistic perspectives, 60

nonterminating, 26, 59, 61, 63, 64

nontotalizable aesthesia, 116

Novarina, Valère, 41

objectivity, 6, 111; constitution of, 78, 118; disinterested, 57; models of, 105, 107, 115, 136; provisional model of, 32; suspension of judgment on, 106; undecidability and, 63; worldly, 178

object-oriented ontology, 2, 118

objects of knowledge: emergence of, 35;

Leibniz, 20; problematic of apprehending, 19
obscurity, 19, 20, 26, 27
On the Gradual Production of Thoughts Whilst Speaking, 76
ontologies of identity, 118
operationally closed, 147, 184
ordinality, 33
organisms, 48, 49, 63, 187; amnesia and, 127; color-spaces of, 153; complex, 131, 144; defining aspects of, 135; environment and, 85–86; as extended critical state of matter, 127; life-defining boundary, 64; memory and, 119; mereological structure of, 135; protomnemotechnical constitution of, 151; routing of inputs, 146; structure of, 126
organism–world system, 35
originary technicity, 137
Origin of Geometry, 136
ostentation, 65–67, 69–70, 72
otherness, 50
overdetermination, 64, 80, 97
Oyama, Susan, 86

Palazzo Enciclopedico, Il (Auriti), 69
Panglossian, 87–88
panpsychism, 1–2, 4, 198
paradise crows, 82
paradoxes, 36, 42, 129, 149
Parmenides, 27, 208
passive synthesis, 32, 124
peacocks, 81–82
Peirce, Charles Sanders, 174
perception, 22; entanglement of little, 201; knowledge and, 44; sensation and, 129; temporal, 130
perceptronium, 128; aesthesis and, 192; computronium and, 197–98; conditions of, 8, 66; describing, 197
performative contradiction, 156, 159
permeability, 141

petites perceptions, 26
Petitot, Jean, 32
Phaedrus (Plato), 140
pharmakon, 139
pharmakos, 53
phase transitions, 118–19
philosophy, 37–38, 46; aesthetics and, 23; Deleuze on, 7; internal pressures of, 32; knowledge and, 11; pre-Socratic explosion of, 28
photography, 11
physics, 33, 164, 169, 171–72, 175. *See also* quantum physics
Plato, 43, 75, 138, 140, 175
Platonic idealities, 32
pluralism, 173–74
plurality, 27, 146
poiēsis, 67
Poincaré, Henri, 30, 57
Poincaré-Berthoz isomorphism, 32
poisedness, 145
polis, 1, 2
Popper, Karl, 9, 165
possibilism, 203
possibilities, 173
postmodernism, 77, 155
poststructuralism, 77, 155
presignifying life-world, 100
presuppositionlessness, 39
presymmetry, 180
Prigogine, Ilya, 110
primary qualities, 29
primary selection principle, 87
principle of least action, 170
prosthesis, 109, 140, 146
protention, 138
protomnemotechnical constitution of organisms, 151
protomnemotechnics, 141
prototechnical adaptations, 144
prototechnics, 150
Prum, Richard, 96, 97
pseudo-randomness, 54, 58, 64

pseudo-technicity, 114
psychological states, 3
psychosocial individuation, 100, 153
pure aestheta, 65
Putnam, Hilary, 214
Pyrrhonistic, 64, 79, 106, 177, 211

qualia, 4, 20, 188–89; cognition and, 191; *haecceity* and, 190; ineffability of, 21, 24, 63; *je ne sais quoi* of, 21; Leibniz and, 59; *petites perceptions*, 26; *qualia*-infinitessimal isomorphism, 86; stability of, 207; synthesis of, 192
quantum computing, 199
quantum physics, 17, 33, 34, 160, 168; democratic nature of, 194; early theory of, 193; Heisenberg and, 176; loop quantum gravity, 175; microscopic, 182; quantized values of, 186; quantum eraser, 179; quantum fluctuation, 183; quantum strangeness, 166; unitarity, 203
quasi-causal(ity), 16, 171, 177, 213; entanglements, 207; necessity and, 203; incursion, 183; mechanism, 182; transition, 189

Ramachandran, Vilayanur, 80, 89, 90
Rancière, Jacques, 68
randomness, 168, 208; electrical fluctuations, 93; generation techniques, 52, 54–57; pseudo-, 54, 58, 64
rasa (essence), 89
rational numbers, 26
real experience, 5
realism, 16, 155, 214
realizability, 60
reason, 1, 6, 29, 174
reasoning, 21, 136
reciprocal inclusion, 117
reductionism, 38
reflective judgment, 49
reflective mentation, 152
Reichenbach, Hans, 34, 160

reincarnation, 139, 140
relativism, 35, 105, 155, 159
relativity, 34
Renaissance, 45
repetition, 124, 132, 134, 228n58
representation, 145
reproduction, 145
resonance, 93
retention(s), 128, 129, 133, 138, 141; mnesic synthesis of, 155; retentional exteriorization, 145
Riegl, Alois, 77
Riemann, Bernhard, 30
Roden, David, 153
Rorty, Richard, 34
Rovelli, Carlo, 175–76
Rule 110 cellular automaton, 58, 61
runaway effects, 92
Ruskin, John, 9, 45
Russell, Bertrand, 9, 38, 60

sapience, 5; human, 1, 152; sentience and, 43, 63, 153
Sartre, Jean-Paul, 68
satisfice, 87–88, 91, 97
savannah hypothesis, 80
scarcity, 82
Schaeffer, Jean-Marie, 102
Schelling, Friedrich Wilhelm Joseph von, 114–15, 167
Schlick, Moritz, 33
Schmidt, Klaus, 72
Schrödinger, Erwin, 160, 193
science, 45, 47, 210
scientific hypothesis, 10
scientific paradigms, 104
Searle, John, 199
secondary integration, 128, 132, 133
secondary qualities, 29
sedentary life, 73, 74
sedimentation, 127
selection, 85, 94; aesthetic, 12, 97, 99; environmental, 71, 93
selective pressures, 95

self-awareness, 198
self-organization, 57, 105, 147, 167
Sellars, Wilfrid, 3, 9, 39, 164
Selme, Leon, 181
Semper, Gottfried, 76–77
Semperian–Rieglian debate, 76, 79
sensation, 4, 5, 42; knowledge and, 28, 30; perception and, 129
sentience, 212; animal, 1, 152; sapience and, 43, 63, 153
serendipitous encounters, 51
serious activities, 68
Serres, Michel, 41
sexual selection, 12, 67, 71; aesthetics and, 96; animal territorializations and, 98; art and, 69, 81; not taken seriously, 83
Shannon, Claude, 78, 165, 169
Shaviro, Steven, 49
signaling theory, 12, 67, 83
Simondon, Gilbert, 79, 100, 110, 111; on aesthetics, 13, 113; on cognition, 126; generalization of concepts of, 114; intraperceptive images and, 190; metastability and, 181; on time, 131
Smolin, Lee, 174–75, 203, 227n23
snowflakes, 150–51
Socrates, 28, 138, 139, 140
Socratics, 27
Souriau, Étienne, 13, 68, 98
spacetime, 33, 58
spandrel, 84, 87
spatium, 117
Spinoza, Baruch, 63
state vector, 160
static genesis, 158, 176, 178, 205
Stiegler, Bernard, 99, 104, 112, 136; on default of origin, 114; exteriorization and, 143; hypomnemata and, 121; on memory, 141; mnemotechnics and, 13; originary technicity and, 137; Plato and, 138
structural coupling, 135, 183, 184, 187, 188

subjectivity, 4, 17, 110, 111, 199; aesthesis and, 207; agential, 63; constitution of, 118; dissociation of subjective act, 51; fixed form of, 7; forms of, 172; human transcendental, 214; hybridizing, 215; incapacity of, 23; subjective necessity, 167; subjectivist idealism, 157; transcendental constraints on, 6; transcendental influences on, 10; types of, 191; world beyond, 18
subject–object relation, 163
sublime, 62, 63
substrate independence, 97, 191, 212
superject, 128, 187, 191, 192
supernormal stimulus, 88–89
superposition, 97, 159–60, 175, 178, 205; absolute, 169; atemporal, 162, 163; coherent state and, 194; consistency and, 213; constantly reinstantiated, 180; indiscriminate, 168; probabilistic, 193; static, 167
supersession, 118, 146, 149; eventual, 114; intercession and, 13, 14, 105, 144, 202, 208; material emergence and, 116; mnemotechnical assemblage and, 130; prosthesic, 178
supplements, 150, 180
surface of translation, 42
surrealism, 46
suspension of judgment, 106
symmetry breaking, 54, 178, 179–80, 205
synthesis, 17, 47, 215; analytic–synthetic distinction, 65, 138; cognition and, 50; general form of, 202; of habit, 141; levels of, 136, 154; of memory, 134; mnesic synthesis of retentions, 155; passive, 32, 124, 125; of *qualia*, 192; retroactive, 138; second, 128, 129, 130, 145
synthetic reason, 29

technē, 67, 99, 101, 111, 211; evolution and, 79
technical being, 127

technical determinism, 11

technics, 110

Technics and Time, 136

technogenesis, 210

technology, 12

technoscience, 209

technotranscendental conditions, 122, 141, 144

Tegmark, Max, 17, 192, 200, 201

Teissier, Bernard, 32, 33

temporal perception, 130

territorialization, 98–100, 103, 111; aesthesis and, 101; deterritorialization, 104, 106

thermodynamics, 126, 168

thing in itself, 36

Third Critique (Kant), 168

three-body problem, 57

Thue, Axel, 62

time, 3, 181; Aionic temporality and, 190; ancestral, 156; Aristotle on, 120; conditions for, 15; Deleuze on second synthesis of, 128, 129; denial of, 175; intelligibility of, 17; internal experience, 144; logical, 176–77, 178, 182, 186; Simondon on, 131; technotranscendental conditions of, 122; temporal perception, 130; temporal realism, 205; time consciousness, 121

Tinbergen, Niko, 88

togetherness, 202

Tononi, Giulio, 17, 192, 195, 196, 198

total science, 106, 115

trace, 38

transcendental aesthetic, 35

transcendental analytic, 33

transcendental empiricism, 5, 34, 180

transcoding, 100

transduction, 86–87, 115–17, 202

transindividuation, 215

trivial complexity, 62

Turing, Alan, 56, 58

Turing universality, 58, 61, 112

Turner, Victor, 103

Tzara, Tristan, 51

Uexküll, Jacob von, 10, 122

Umwelt, 7, 47–48, 89, 105, 122; pan-, 206

unconscious inference, 169

uncontroversial centers, 94, 95

uncountability, 10

undecidability, 56, 58, 63, 97, 102

underdetermination, 94, 96–97, 148

Unger, Roberto, 174, 175

unilateral determination, 33, 85, 185

universal complexity, 62

unpredictability, 53

use value, 87

Van Orman Quine, Willard, 9, 34, 47, 94; "field of force," 48; logical positivism and, 38

Varela, Francisco, 86

Veblen, Thorstein, 68

Venice Biennale, 69

virtual, 131, 204

virtual computing, 112

visual representation, 46

vitalism, 4, 211

vital materialism, 2

Voevodsky, Vladimir, 60

Wallace, Alfred Russell, 12, 83, 96

wasps, 86–87

wave function, 160

waves, 193

Wearing, Clive, 133–34, 158

web of beliefs, 10

Weinberg, Steven, 164, 166

Western history of ideas, 50

"What the Tortoise Said to Achilles," 36, 171

Wheeler, John, 16, 78, 165, 166, 179

Whitehead, Alfred North, 14, 60, 114, 117, 189; on actual occasion, 152; eternal objects and, 188

Wilde, Oscar, 78

Wittgenstein, Ludwig, 9, 41
Wolfram, Stephen, 58, 59
world-models, 48, 54, 104, 126, 131, 177, 179
Worringer, Wilhelm, 77
writing, 139, 140

Zahavi, Amotz, 83
Zeno, 36, 117
Zurek, Wojciech, 194

(continued from page ii)

30 The Universe of Things: On Speculative Realism
Steven Shaviro

29 Neocybernetics and Narrative
Bruce Clarke

28 Cinders
Jacques Derrida

27 Hyperobjects: Philosophy and Ecology after the End of the World
Timothy Morton

26 Humanesis: Sound and Technological Posthumanism
David Cecchetto

25 Artist Animal
Steve Baker

24 Without Offending Humans: A Critique of Animal Rights
Élisabeth de Fontenay

23 Vampyroteuthis Infernalis: A Treatise, with a Report by the Institut Scientifique de Recherche Paranaturaliste
Vilém Flusser and Louis Bec

22 Body Drift: Butler, Hayles, Haraway
Arthur Kroker

21 HumAnimal: Race, Law, Language
Kalpana Rahita Seshadri

20 Alien Phenomenology, or What It's Like to Be a Thing
Ian Bogost

19 CIFERAE: A Bestiary in Five Fingers
Tom Tyler

18 Improper Life: Technology and Biopolitics from Heidegger to Agamben
Timothy C. Campbell

17 Surface Encounters: Thinking with Animals and Art
Ron Broglio

16 Against Ecological Sovereignty: Ethics, Biopolitics, and Saving the Natural World
Mick Smith

15 Animal Stories: Narrating across Species Lines
Susan McHugh

14 Human Error: Species-Being and Media Machines
Dominic Pettman

13 Junkware
Thierry Bardini

12 A Foray into the Worlds of Animals and Humans, *with* A Theory of Meaning
Jakob von Uexküll

11 Insect Media: An Archaeology of Animals and Technology
Jussi Parikka

10 Cosmopolitics II
Isabelle Stengers

9 Cosmopolitics I
Isabelle Stengers

8 What Is Posthumanism?
Cary Wolfe

7 Political Affect: Connecting the Social and the Somatic
John Protevi

6 Animal Capital: Rendering Life in Biopolitical Times
Nicole Shukin

5 Dorsality: Thinking Back through Technology and Politics
David Wills

4 Bíos: Biopolitics and Philosophy
Roberto Esposito

3 When Species Meet
Donna J. Haraway

2 The Poetics of DNA
Judith Roof

1 The Parasite
Michel Serres

ALEXANDER WILSON is a Canadian researcher in philosophy of science, technology, and aesthetics, as well as cultural studies, environmental humanities, and media theory. He has held a postdoctoral research position in communication and culture at Aarhus University, Denmark, as well as an assistant professorship in intermedia at Concordia University in Montreal, Canada, and is currently affiliated with the Institute of Research and Innovation, Centre Pompidou, Paris, France. Also an accomplished artist, he has created installations, films, music, and works for the stage. He lives and works in Berlin, Germany.

Milton Keynes UK
Ingram Content Group UK Ltd.
UKHW010629160324
439502UK00012B/352